T0323911

Introduction to Mathematical Models in Operations Planning

Halit Alper Tayalı

Assistant Professor
Istanbul University School of Business
Avcilar, Istanbul

CRC Press
Taylor & Francis Group
Boca Raton London New York

CRC Press is an imprint of the
Taylor & Francis Group, an **informa** business
A SCIENCE PUBLISHERS BOOK

First edition published 2024
by CRC Press
2385 NW Executive Center Drive, Suite 320, Boca Raton FL 33431

and by CRC Press
4 Park Square, Milton Park, Abingdon, Oxon, OX14 4RN

CRC Press is an imprint of Taylor & Francis Group, LLC

Library of Congress Cataloging-in-Publication Data (applied for)

ISBN: 978-1-032-19199-7 (hbk)
ISBN: 978-1-032-19201-7 (pbk)
ISBN: 978-1-003-25809-4 (ebk)

DOI: 10.1201/9781003258094

Typeset in Times New Roman
by Radiant Productions

Preface

Our lives are shaped by unforeseeable events and circumstances that can derail even the most meticulously devised plans. The significant gap between our plans and the realities we face is often caused by economic conditions. In these complex relations between life, plans, and the economy, each aspect influences the others. Understanding these connections, as well as the broader economic context which we live in, can help us in making more informed and effective plans.

Economy refers to the administration and planning of resources within an economic unit, such as a family, community, or nation. In everyday life, it refers to the intricate network of related activities that make up an economic unit, such as production, consumption, or commerce. The term economy is derived from the Greek word 'oikonomia', where 'oikos' denotes a home, and 'nomos' refers to concepts such as law, custom, order, control, regulation, or management. As a result, economy means house rules, and it is no coincidence that running a household is just like running a business.

Briefly, economy is the regulation of resources and economic units should rationally strive to make the best possible, most effective, and efficient use of the resources available to them. Thus, the idea of planning comes into play. Scientific advancements shape modern industrial planning activities. The cornerstone of our most recent industrial revolution, artificial intelligence, has enabled machines to sense, learn, and act. They simplify complex tasks that we find difficult to complete. The economic ideas of demand and supply, however, still apply to businesses: A company's performance depends on how well it balances its supply and demand. In other words, it is critical to strike the best balance between supply and demand to sustain a business.

A company has three core functions: production, finance, and marketing. Production is central to any economic unit and in generating revenue. It significantly differs from the other functions as it transforms concepts into marketable goods or services. Thus, productivity and efficiency in production systems and shop floors is a vital subject.

This book not only outlines the elementary mathematical methods of operations planning with numerical examples but also smoothly introduces the production function of a company to the readers. Furthermore, the book gives detailed information on the R computer programming language along with introductory examples and code scripts, which should be useful for readers interested in operations planning.

Contents

Chapter 1
Introduction

Humans have an extraordinary ability to predict the future, even when they are unaware of it. The physiological nature of how we perceive our universe is unknown, yet, but a healthy individual is a forecaster (Tetlock and Gardner, 2015). Krajewski et al. (2016) and Heizer et al. (2017) define forecast as the art and science of future event prediction for the purpose of planning. Everett and Ebert (1986) distinguish between prediction and forecast: A forecast requires quantitative modeling, and a prediction involves ability, experience, and judgment based on a forecast. Nevertheless, planning, perception, prediction, and forecasting are end products of an estimation process where rough calculations are involved.

Objective and subjective difficulties in modeling the perceived reality might cease to exist for machines can plan better than humans (Tayalı, 2021). We have manufactured and produced to adapt to nature and sustain life. There is no doubt that advances in science, technology, and manufacturing have increased life expectancy. The concepts of demand and supply are central to our lives. Not only does their relationship affect each other, but they can also be inert. The need for cognition fuels this entire process, but there are always a limited number of production factors and there is competition.

The theory of economics examines the interdependent ties between individuals or organizations using the flow of money, products, and services, as well as different subjects like interest rate, inflation, unemployment, and growth. In economics, planning usually implies government intervention: The literature has extensive discussions on how states, either directly or indirectly, intervene in operational activities of economic units (Kazgan, 2000). Scientific advances shape the modern elements of production planning. The first industrial revolution took place in the 18th century. During the same period, Adam Smith saw the enormous effect of division of labor on productivity (Thornton, 2014). It is important to note that productivity and efficiency are related concepts that assess various aspects of performance. Productivity measures the output produced per unit of input, whereas efficiency measures how effectively resources are used to attain a specific goal. A productive process, for example, produces more output with the same input, whereas an efficient process produces the same output with less waste.

Around 1920, the concept of a moving production line revolutionized manufacturing, and during the 1930s, the focus shifted to quality control from a statistical standpoint. By the 1960s, lean manufacturing techniques were introduced. The idea of enterprise resource planning software was developed in the 1970s, and following that, in the 1980s, the theory of constraints emerged as an innovative approach to bottleneck scheduling (Taylor and Russell, 2011). Continuous improvement programs, such as six-sigma, just-in-time, total quality, or business process reengineering implement these revolutionary approaches to production planning and control activities. These conceptual frameworks include a set of standard approaches for analyzing a company's manufacturing processes. However, the problems and requirements of each unique production environment differ.

As a result, businesses should weigh the benefits and drawbacks of these programs and prefer a tailored approach based on their specific problems and needs.

To stay in business, a company must have three core, mainline functions: finance, marketing, and production (Krajewski et al., 2016). Finance is concerned with the acquisition and allocation of scarce resources needed for production. Note that the best resource allocation is a critical decision to make not only within a manufacturing system, but also across a macroeconomic unit. Because it converts physical materials or abstract ideas into goods or services, a company's production function is the foundation of its financial performance, while the finance function, another core function of the company, supports real investment in production and operations (Zhao and Huchzermeier, 2015). Finally, the marketing function generates revenue through the sale of manufactured goods or produced services. It is worth noting that the intersections of these three core functions offer exciting opportunities for future research. For instance, Cachon and Terwiesch (2012) have reported the interaction between the operations of a company and its financial performance by underlining that the key driver of a company's financial performance is its operational variables.

With the Fourth Industrial Revolution, the world is undergoing a digital transformation, and globalization is hastening the pace of this transformation. There are discussions about the deteriorating effects of globalization on, for instance, the distribution of income, or global warming, but the industrial paradigm keeps shifting. The hallmark of this revolution is artificial intelligence: machines that can sense, learn, and act. As a result, the current operational practice is to quantify and digitize data, as well as automated and optimized processes, so that machines can work quickly on massive datasets and drop human inefficiency. Companies require insightful data and analysis to compete in the market, just as a car requires oil and gas to run. Mathematical models enter the picture at this point.

A mathematical model is a simplification of a real-world, complicated system that describes how its inputs lead to its outputs. Analytical modeling and quantitative approaches are the emphasis of operations research, which is the study of operational problems using mathematics and statistics. Business analytics, on the other hand, is the study of data and quantitative modeling to understand the performance of a company. It focuses on the development of strategies via the use of technology, skills, and applications to get value from data. By analyzing data with a variety of tools and techniques, ranging from descriptive statistical approaches to advanced data mining, one can extract that valuable information.

The link between a company's supply and consumer demand figures out its core performance. The goal of running a business is to match supply and demand as optimally as possible (Cachon and Terwiesch, 2012). Although a company's production planning method should include all core and supporting functions, companies should dive into the production function and continuously look for ways to collect data, model, and automate operational processes. The level of digitalization of production processes is critical for obtaining data and feeding it into mathematical models of operations planning.

Operational planning and control activities oversee a variety of organizational difficulties and problems. The responsibility of a production planner is to plan to manage an operation, and coordinate business activities with lucrative, and ideal choices. A production planner should prefer to quantitatively model production processes to reduce operating costs. Integrating a production planning and control system into a business, on the other hand, is a difficult problem. It is possible that a production planner will not even know where to start, but the important thing is to start out with a broad vision. By extensively discussing production and operations planning, Chapter 2 tries to acquaint readers with current scientific language in the field.

One can think of the production environment as either a system or an entity, made up of components that work together to achieve a particular goal. The two main categories of manufacturing processes are job type and flow type. In-depth details, analysis, and numerical examples on processes taking place in a manufacturing environment are in Chapter 3.

The problem of facility layout refers to the physical arrangement of manufacturing operations in relation to one another. There are four hypothetical layout strategies that position manufacturing processes and operations in relation to each other: Manufacturing cell, assembly line, project, and work center types. However, it is usually the workflow pattern which dictates the layout of a manufacturing environment. Consequently, a mix of layout types at a site should not alarm a production planner. Details and illustrations about the facility layout problem are in Chapter 4.

A project refers to a set of linked tasks or workpieces that are restricted by and dependent on predetermined production factors. A methodical approach might assist a production planner in managing a project more efficiently. The goal of project planning is to estimate the completion date for monitoring and regulation purposes. For an improved production process, critical path method figures out the scheduling information for each activity within a project. In Chapter 5, numerical examples of the critical path technique for project planning and scheduling are provided.

A job shop manufacturing environment exists in lots of facilities or factories around the world, producing small quantities of output with extremely variable product characteristics. Making the best operations' scheduling decision in these environments is a difficult assignment for the production planners since, depending on the physical structure or the configuration of the manufacturing environment, a planner might face different barriers.

Chapter 6 attempts to describe and summarize operations scheduling problems with different heuristics and assumptions. The primary fiscal goals of a production planner are to minimize costs or maximize profits. The field of optimization theory is concerned with the critical points or values for a mathematical function, such as the optimum values and the extremum or saddle points, if any exist. An optimization model, for example, could quantitatively solve for the optimal mix of production levels for a set of products produced at a facility. Based on a set of underlying and, preferably, simplifying assumptions, such an optimization model can solve for the maximum expected value of total profit or the minimum expected value of total cost throughout the entire production process.

A linear optimization model has only linear functions, and in a commercial setting, its constraints are often related to a company's production factors and available resources. Chapter 7 covers how to transform a manufacturing scenario into the language of optimization modeling. The use of quantitative techniques from operations research extends beyond a company's core function of production to other core functions such as finance or marketing. Furthermore, a planning researcher could run into various problems that require optimization knowledge, not just in practical economics contexts but also in theoretical economics.

Aggregate operations planning helps a company in figuring out the quantities of various production factors for a predefined planning horizon. There are various quantitative methods for overcoming the difficulties of the aggregate operations planning problem, as well as graphical and spreadsheet models. The development of an optimization framework for the aggregate production planning problem sparked a revival in production management in the 1960s: The linear optimization model for the aggregate operations planning problem presents a thorough solution with its objective function of cost minimization. The book presents a detailed explanation of the planning model in Chapter 8, which provides a strategic tool for various time periods of the planning horizon. Moreover, it includes a brief documentation of the solution using the R computer programming language.

References

Cachon, G. and Terwiesch, C. (2012). *Matching Supply with Demand: An Introduction to Operations Management.* McGraw-Hill/Irwin.

Everett, A. and Ebert, R. (1986). *Production and Operations Management: Concepts, Models, and Behavior* (3rd Edn.). Prentice-Hall.

Heizer, J., Render, B. and Munson, C. (2017). *Operations Management: Sustainability and Supply Chain Management* (12th Edn.). Pearson Education Limited.

Kazgan, G. (2000). *Economic Thought (in Turkish)* (9th Edn.). Remzi.

Krajewski, L., Malhotra, M. and Ritzman, L. (2016). *Operations Management: Processes and Supply Chains* (11th Edn.). Pearson.

Tayalı. (2021). Demand Forecasting Models with Time Series and Random Forest. pp. 76–99. *In*: A.M. and A.A.S. Bushell (eds.). *Driving Innovation and Productivity Through Sustainable Automation.* IGI Global. https://www.igi-global.com/chapter/demand-forecasting-models-with-time-series-and-random-forest/274150.

Taylor, B.W. and Russell, R.S. (2011). *Operations Management: Creating Value along the Supply Chain.* John Wiley & Sons.

Tetlock, P. and Gardner, D. (2015). *Superforecasting: The Art and Science of Prediction.* Crown.

Thornton, P. (2014). *The Great Economists: Ten Economists whose Thinking Changed the Way We Live.* Pearson Education Limited.

Zhao, L. and Huchzermeier, A. (2015). Operations-finance interface models: A literature review and framework. *European Journal of Operational Research*, 244(3): 905–917. https://doi.org/10.1016/j.ejor.2015.02.015.

CHAPTER 2
Operations Planning

The primary foci of research conducted and published in the discipline of operations management are the design, acquisition, development, and delivery of products and services. In the body of research about operations planning, one is highly likely to come across the following three terms: process, operations, and supply chain. The definitions are included in the chapter, which aims to familiarize readers with current scientific jargon in the field by thoroughly describing production and operations planning.

A planning activity is conducted to manage an operation. Operational planning and control activities cover and are expected to solve a wide variety of organizational difficulties and problems. A production planner aligns the organization's operations by coordinating the activities involved in producing goods and services with profitable decisions. To lower operating costs, the production planner is typically expected to quantitatively model the production processes. In operations planning, decision analysis models are often used as mathematical representations of quantitative business processes for analyzing decision alternatives and making better decisions.

In operations planning, tasks are classified in a variety of ways, the most common of which is by time. Long-term operations planning and control, for example, are focused on yearly planning horizons, while short-term and intermediate-range tactical planning and control decisions are focused on daily or weekly planning horizons.

Integrating a production planning and control system is a challenging problem for businesses because it requires the development of a comprehensive, data-centric point of view. Monitoring performance levels, matching supply and demand, aggregate planning, distributing jobs to people or machines, material requirements planning, sequencing, and scheduling jobs, and other shop floor control activities are some of the planning responsibilities of the operations staff, who are typically in charge of tactical and operational decisions that utilize a facility's production function. A production planner may even be unsure where to begin, and the key point is to begin with a broad perspective. To make things easier, trying to understand the level of data digitization in the production function is beneficial because it is difficult to analyze, plan, automate, or improve manufacturing processes without digital data.

In terms of revenue generation, the company's production function differs significantly from the other functions. According to Cachon and Terwiesch (2012) and Krajewski et al. (2016), businesses will gain from using operations planning tools since an efficient operational plan and process design may produce wealth and value for the company. Heizer et al. (2017) supply compelling evidence to support this claim by showing that, of the three main alternatives listed below, lowering production costs contributes the most to net profit:

- Boost sales revenue by 50%
- Reduce financial costs by 50%
- Reduce production costs by 20%

When compared to the other options related to increased sales or decreased financial costs, a minor reduction in production costs makes the greatest contribution to the net profit (Tayalı, 2021).

Production planning and control activities encompass a wide range of concepts and are expected to offer solutions to a wide range of organizational issues or problems. Setting goals, organizing teams, preparing bills of materials, forecasting demand, assigning labor, measuring quality, and scheduling operations are examples of these activities. All these ideas contribute to the goal of operations management, which is to match supply and demand (Cachon and Terwiesch, 2012; Zhao and Huchzermeier, 2015).

These three terms appear often in the literature on production planning: process, operations, and supply chain. A process is a collection of activities or tasks that convert inputs to outputs. An operation is a collection of resources that conduct or perform processes. It is limited to an action on the product or service. It is important to keep in mind that production is an all-encompassing concept for an operation. Finally, a supply chain is the collection of operations that results in the production of a service or a product. Again, it should be recognized that these terms are nested within each other, and production is a catch-all term. A manufacturing environment is full of operations that convert abstract thinking and physical substances into finished goods and finished services or products.

The implementation of a production planning and control tool is a challenging task for the company, which demands a multidimensional perspective. To reduce operational costs, one should mathematically model the manufacturing processes. These mathematical models that reflect production activities do not always have to be novel or complex. Even a mathematics professor takes pride in not using mathematics (Tetlock and Gardner, 2015). Because there are so many mathematical ideas and approaches in the production planning and control literature, it is impossible to study all the methods or variables to choose a single approach; instead, a broad perspective combined with reasoned judgment should suffice.

Operational research and operations management

All economic units should act to produce goods or services efficiently. Planning is used to manage operations and align supply with demand. This link between supply and demand refers to a broad range of decisions about design, procurement, quality of goods or services, strategies of process, location, layout, inventory, as well as scheduling, and maintenance, among other aspects. The field of production and operations management helps companies make profitable and sustainable choices by organizing these decision-making mechanisms. In short, operations management is concerned with the management of the operational activities of goods and services, whereas operations research is concerned with mathematically modeling a problem and attempting to find an optimal solution (Fuller and Martinec, 2005). Concisely, the common ground between the two areas is their emphasis on input–output transformation for the purpose of making sustainable decisions with limited resources. Eventually, it turns out that the probability that these choices will be wise ones is typically increased by mathematics.

Mathematics is a broad and diverse field that studies numbers, quantities, shapes, and related topics. It is softly divided into two interdependent branches of pure and applied mathematics where the applied branch typically studies natural phenomena. It includes topics such as statistics, and quantum mechanics. Pure mathematics, on the other hand, studies abstraction, and topics such as algebra, trigonometry, and calculus (Nelson, 2008). The word mathematics is derived from the Greek word 'mathema', which means "that which is learned, learning, science" (Klein, 1971). Each mathematical branch has its own set of tools and methods for creating abstractions, modeling phenomena, and solving problems. Many of these fields frequently overlap and intersect enabling researchers to approach complex issues that require more than a single point of view. Thus, by exploring various branches of mathematics, one can discover novel solutions to problems.

Operations research is a field that studies mathematics to address operational issues. It is also referred to as the science of better. Most of the time, these problems are resolved through a decision-making process. Thus, operations research entails applying scientific models, primarily mathematical and statistical ones, to decision-making problems. According to Farahani et al. (2010), every optimization problem is a decision-making process with two outcomes: yes or no. Typically, optimization problems involve resource allocation. Many operational problems seek solutions to cost minimization, profit maximization, task scheduling, and inventory policies. Some of these problems may contain stochastic variables, and thus require statistical analysis. Linear, nonlinear, or dynamic programming, critical path analysis, game theory, and simulation are just a few of the approaches and techniques used to solve problems in operations planning (Kandiller, 2007; Nelson, 2008; Srinivasan, 2017; Tayalı, 2016).

Planning tasks

Managers of economic units must make the best decisions possible to compete and survive in the market. A production planner's responsibility is to match supply with demand and to create connections to ease this process. Demand and supply definitions are deeply rooted in economics. Managers or researchers should clear up any misunderstandings so that everyone speaks the same language, because the roles, attributes, literal meanings, or definitions of the actors in this chain of demand and supply may differ depending on the direction of the trade relationship. External customers generate demand by purchasing a company's goods or services, while internal customers require input from the supply side to achieve their production-related goals. The term 'clients' is sometimes used to refer to internal customers within a supplier. An external supplier provides goods or services from outside of the organization, whereas an internal supplier provides goods or services from within the organization.

A production planner's responsibilities are extensive. Multiple concepts from business, mathematics, and engineering must be thoroughly understood. Furthermore, knowing how to use a production planning and control tool is essential for saving time and energy. Simply put, a production planning tool should consider the production factors. These factors are broadly capital and labor, or the inputs to the well-known Cobb-Douglas production function (Kazgan, 2000; Pindyck and Rubinfeld, 2000). These production factors have also been expanded upon: leader, manager, worker, machinery, equipment, facility, material, land, energy, entrepreneurship, ability, and method of production. For exceptional circumstances, the list of production factors may be expanded. On the other hand, the output of a company's production function can be a tangible good, an intangible service, or a combination of the two, depending on the nature of the business.

The economic value of a company is created via the company's production and operations and the scope of planning related tasks is vast. Due to the plethora of obstacles involved, it is impossible to define these planning tasks with mathematical precision. Consider the distinction between the definitions of manufacturing and production: A manufacturing process typically produces a tangible output, while a service process typically produces an intangible output. This raises multiple issues, such as whether a car or a software is the result of a series of manufacturing or service processes. The answer is that the manufacturing and production processes are included in both the car as a tangible product and the software as an intangible service. It is worth noting that there have been countless attempts in literature to specify the differences between manufacturing and service processes and to supply common definitions for factors of production.

Before describing and classifying production planning tasks, it is crucial to stress the importance of data. Access to quantitative and digital data is essential for effective production planning, automation, optimization, and generating valuable insights. The decisions made by a company's operations or production function can be grouped into three temporal categories based on their planning horizon (Chase et al., 2006; Heizer et al., 2017; Jacobs and Chase, 2018; Krajewski et al., 2016; Nahmias and Olsen, 2015; Tayalı, 2021):

- Long-term strategic choices
- Intermediate-term tactical decisions
- Decisions on short-term operational planning and control

Long-term decisions, such as facility location, are irreversible and expensive to reverse once made. Top executives are usually in charge of these long-term strategic decisions, which should span years or decades. Strategic decisions may be related to research and development, supply planning, inventory positioning and figuring out decoupling points, product development, and changes and investments related to production factors. Market research, product development, demand forecasting, and human resource planning are all examples of long-term production or operations planning and control activities.

Intermediate-term tactical and short-term operational plans concentrate on planning enterprise resources within a predefined planning horizon. Months, weeks, days, hours, minutes, and even seconds can be used to define the planning horizon. The operations staff oversees making tactical and operational decisions that make use of the facility's production. The planning duties of the operations staff include monitoring performance levels, matching demand with supply, aggregate planning, assigning jobs to workers or machines, planning material requirements, sequencing and scheduling jobs, and other shop floor control activities. To increase profitability and supply better services to society, these operational decisions often involve quantitative modelling (Tayalı, 2021).

The aim of this book is to examine fundamental planning models in a manufacturing job shop, so it does not delve into the definition of logistics. However, it is important to acknowledge that there is a discussion regarding the origin of the term, which is typically used to refer to the movement and transportation of goods. Note that formal language may use the terms planning and logistics interchangeably.

The following chapter describes manufacturing processes, including a classification of processes, detailed information on the nature of job shops, and process analysis examples.

References

Cachon, G. and Terwiesch, C. (2012). *Matching Supply with Demand: An Introduction to Operations Management*. McGraw-Hill/Irwin.

Chase, R., Robert, F. and Nicholas, J. (2006). *Operations Management for Competitive Advantage* (11th Edn.). McGraw-Hill/Irwin.

Farahani, R.Z., SteadieSeifi, M. and Asgari, N. (2010). Multiple criteria facility location problems: A survey. *Applied Mathematical Modelling*, 34(7): 1689–1709. https://doi.org/10.1016/J.APM.2009.10.005.

Fuller, J. and Martinec, L. (2005). Operations research and operations management: from selective optimization to system optimization. *Journal of Business & Economics Research*, 3(7): 16.

Heizer, J., Render, B. and Munson, C. (2017). *Operations Management: Sustainability and Supply Chain Management* (12th Edn.). Pearson Education Limited.

Jacobs, F.R. and Chase, R.B. (2018). *Operations and Supply Chain Management* (15th Edn.). McGraw-Hill/Irwin.

Kandiller, L. (2007). *Principles of Mathematics in Operations Research*. Springer.

Kazgan, G. (2000). *Economic Thought (in Turkish)* (9th Edn.). Remzi.

Klein, E. (1971). *A Comprehensive Etymological Dictionary of the English Language: Dealing with the Origin of Words and Their Sense Development Thus Illustrating the History of Civilization and Culture* (Unabridged). Brill Academic Publishers.

Krajewski, L., Malhotra, M. and Ritzman, L. (2016). *Operations Management: Processes and Supply Chains* (11th Edn.). Pearson.

Nahmias, S. and Olsen, T. (2015). *Production and Operations Analysis* (7th Edn.). Waveland Press.

Nelson, D. (2008). *The Penguin Dictionary of Mathematics*. Penguin UK.

Pindyck, R.S. and Rubinfeld, D.L. (2000). *Microeconomics* (5th Edn.). Pearson Education (US).

Srinivasan, G. (2017). *Operations Research: Principles and Applications* (3rd Edn.). PHI Learning.

Tayalı, H.A. (2016). *Statistical Variance Procedure Based Analytical Hierarchy Process: An Application on Multicriteria Facility Location Selection* [Doctoral thesis, Istanbul University]. https://tez.yok.gov.tr/UlusalTezMerkezi/.

Tayalı. (2021). Demand forecasting models with time series and random forest. pp. 76–99. *In*: A.M. and A.A.S. Bushell (eds.). *Driving Innovation and Productivity Through Sustainable Automation*. IGI Global. https://www. igi-global.com/chapter/demand-forecasting-models-with-time-series-and-random-forest/274150.

Tetlock, P. and Gardner, D. (2015). *Superforecasting: The Art and Science of Prediction*. Crown.

Zhao, L. and Huchzermeier, A. (2015). Operations-finance interface models: A literature review and framework. *European Journal of Operational Research*, 244(3): 905–917. https://doi.org/10.1016/j.ejor.2015.02.015.

CHAPTER 3
Manufacturing Processes

The manufacturing environment can be thought of as either a system or an entity, since it is made up of components that have been brought together to work toward a certain goal. This chapter presents in-depth information, analysis, and numerical examples related to processes ongoing in a manufacturing environment.

Factors of production are the input to a company's production function. Operational processes, on the other hand, refer to the transformational stage. The outputs are the final goods or services. Although every link in this chain is crucially vital, the operational processes — the procedures or the actions that convert inputs into outputs — might be the most fundamental one.

The operational processes of sourcing, manufacturing, and delivering are the three primary focuses of a production plan, whether they are conducted by humans or machines. It is possible to categorize a manufacturing process or environment in a variety of diverse ways, depending on factors such as the process type, the raw materials used, the product flow, and the product volume. For instance, to produce a physical output, a manufacturing process needs to source and store components, as well as other relevant factors of production. These stored components are also known as stocks or inventories of materials. The level of inventory might be a way to classify a manufacturing environment. Manufacturing processes can also be broken down into two primary categories: job type and flow type. The movement of a product from one stage of production to the next is referred to as 'flow' in the context of the manufacturing process. Consequently, a production planner can see two types of operations within a manufacturing environment: A process-oriented operation or a product-oriented operation. These various sorts of classifications that are used in manufacturing are simplistic attempts to describe the relationship between the process and the product. On the other hand, it is beneficial for the production planners to name a manufacturing process so that they might expect probable operational problems, such as bottlenecks.

As the collection of actions that convert inputs to outputs, processes are at the core of production. Companies use manufacturing as a competitive advantage and invest to improve process efficiency, product quality, and customer satisfaction. The action of a human or a machine on a product figures out the scope of an operation in a manufacturing setting. The manufacturing environment may be considered as a system or an entity that is formed of components brought together to conduct a shared aim.

The abstract components of a system are identified as its input, transformation phase, output, and feedback process. Factors of production are the inputs of a manufacturing system, operational processes are the transformation phase, and commodities or services are the resultant output. Feedback is a monitoring tool that decides if the transformation phase is functioning properly.

Analytical thinking entails breaking something down or splitting it into its constituent pieces or concepts. That is where the term analytics receives its meaning. Each system could be a part of a greater system, and it also might make up multiple subsystems. For instance, a moving line might be a part of an installation system which might be a part of a facility. That facility might be a part of a company which might be a part of a supply chain. Finally, that supply chain might be a part of a macroeconomic unit.

While the focus of a manufacturer or supplier is production, there are three crucial tasks involved in operations management: procurement (sourcing), production (making), and distribution (delivering). Companies may choose to use a manufacturing execution system to watch and manage related activities. These information systems are software that allows one to track and record what happens in a production system. Additionally, this software may offer choices for perfecting operations using integrated analytical tools, as well as supplying real-time interfaces to other software used by equipment, personnel, customers, or suppliers inside or outside the facility (Tayalı, 2021b; Turban et al., 2014).

A process can be broken down into three abstract components: task, flow, and storage. Tasks refer to the specific activities that need to be performed to complete the input–output transformation, either in part or in full. The movement of a product from one task or an activity to the next one is referred to as flow within a manufacturing process. Finally, a manufacturing process requires storage, or stocks or inventories of materials, and resources to produce a good.

There are different ways to categorize a manufacturing process or a manufacturing environment, such as by process type, raw material, product flow, and product volume.

Different types of manufacturing environments require different layout analysis methods. To achieve good productivity, a company needs to use the right layout analysis techniques for organizing its operations within a manufacturing environment. It is also helpful for the production planner to name a manufacturing process to prepare for operational difficulties. In the end, these categorizations are simple logical attempts to describe the process-product link. However, figuring out the sort of process or manufacturing environment that a company has is difficult since companies often show a variety of manufacturing processes, strategies, or environments inside their facilities. Furthermore, one should keep in mind that these categories might always be fuzzy, meaning that a process may not be a complete member of a single class but display the broad traits of another.

To begin with a universal categorization of a manufacturing process, there are two main types: job and flow (Chase et al., 2006; Kasapoğlu and Tayalı, 2012). They are also known as job shop and flow shop, when referring to the production environment. Table 1 compares the main characteristics of these two categories.

Table 1. Features of job and flow types of manufacturing processes.

	Job	Flow
Material handling and movement	High	Low
Product type variation	High	Low
Variability of routing precedence	High	Low
Variability of demand	High	Low
Inventory of raw material	Low	High
Inventory of work-in-process	High	Low
Inventory of completed goods	Low	High
Required artisanship	High	Low
Ease of planning, monitor, and control	Low	High
Production volume	Low	High
Machine and equipment utilization rate	Low	High

In a job shop, items are transported in an intermittent way, whereas in a flow shop, they are transported consistently, usually via a conveyor belt. Work-in-process storage is used to hold a product until it is ready for the next stage in its production process when it is not being processed or transported. This type of storage is more common in job shops than in flow shops. A job shop typically features universal machinery and equipment to create a wide range of products, but a flow shop has unique gear and equipment intentionally constructed or planned to produce a narrow range of goods. As a result, various product pathways may be seen in a job shop, but all goods in a flow shop may follow a single route. In job shops, machinery, equipment, or personnel often function in parallel, but in a flow shop, they follow a sequence. Therefore, although a work shop's rate of utilization might be adjusted simply by adding or removing inputs, a flow shop's utilization rate is not as responsive.

A manufacturing environment may be easily named by a production planner based on its interactions with inventories and the product lead time, which is the time it takes to deliver a product. Considering these two characteristics, manufacturing environments or strategies may be classified into four distinct types:

- Make-to-stock
- Assemble-to-order
- Make-to-order
- Engineer-to-order

Make-to-stock refers to a production environment or strategy in which a company stockpiles items that can be supplied at once. This sort of production environment should allow the organization to supply the items as soon as possible. However, it demands the most expenditure for inventory management compared to the other types of strategies. When a company installs preassembled components or subassemblies for a finished product, the production environment is using an assemble-to-order method. If, on the other hand, the company constructs items directly from raw materials and components, the production environment for that product is considered to follow a make-to-order approach. The production procedures in all three methods adhere to the product specifications. In an engineer-to-order (or a design-to-order) manufacturing environment, however, the company collaborates with the customer to design the final product and construct it using raw materials and components. When compared to the other strategies, this sort of production environment has the longest customer lead time and the smallest amount of inventory investment.

Product and process relationships: Operation types

In addition to making strategic choices concerning the manufacturing environment, a company chooses the type of process it will use to plan and execute its operations. In manufacturing environments, a production planner may see two kinds of activities: process-oriented operations and product-oriented operations. These types of activities may be valid across the whole production plant or simply in a specific operations area.

A process-oriented operation is referred to as a job shop, a process-focused facility, or a work center in the manufacturing literature. Nonetheless, the emphasis of a process-oriented operation is on the process rather than the product or the output. This shows that the product may take an intermittent path and wait for the next step in the operation after it has been processed by a different machine, since the other resource of the same operation may be occupied with another activity. Briefly, process-oriented facilities are arranged according to their processes. Companies that dedicate resources for process-oriented activities include cabinets, paint, print, or machine shops, restaurants, and foundries. These companies specialize in custom-made, tailor-made or prototype items, specialized or sophisticated instruments, and industrial gear, equipment, and products. The product diversity is exceptionally broad since these companies concentrate their efforts on their process capability rather than on designing a specific product.

In contrast, a product-oriented operation is one that revolves around a certain product. A product-oriented operation is a continuous manufacturing process that transforms raw materials into completed goods, often all at once. This sort of operation may be referred to as an assembly line, continuous flow, flow process, or mass customization by the production planner. The primary notion behind a product-oriented operation is that resources are committed to a specific product, as opposed to a process-oriented operation, in which firm resources are assigned to manufacture a wide range of goods. Product-oriented operations are used by facilities that want to focus on manufacturing a specific kind of product. Chemical manufacturing, for example, often requires product-oriented activities.

Bottlenecks

A bottleneck can refer to various aspects within a production system, such as a process, a step within a process, a machine, a work center, or an employee. It can be described as a production factor that operates at a lower rate than demand (Chase et al., 2006), which restricts the system's ability to function at its maximum output. In other words, it is the process with the highest workload or the longest processing time per unit. To improve efficiency, it is important to minimize nonproductive time at the bottlenecks and focus automation efforts on these processes. Bottlenecks are useful in planning and monitoring operations and are optimal spots for controlling costs and production flow (Taylor and Russell, 2011). For instance, the theory of constraints, a continuous improvement program developed by Goldratt (1980), uses bottlenecks as the key component of operations scheduling (Tayalı, 2021a, 2016).

Example 1

A bottleneck within a manufacturing system might be found by examining the workloads of workstations inside that system. This and the next example are simplified representations of the cases given at Krajewski et al. (2016) which explain the nature of bottlenecks that can arise at a manufacturing setting where there exist several products. Suppose a procedure generates two different products, A and B. Product A visits processes 1 (4), 2 (5), 3 (11), 4 (14), and 8 (9) whereas Product B visits processes 1 (4), 2 (5), 5 (4), 6 (19), 7 (11), and 8 (9), with the numbers within the parenthesis indicating the length of each product in minutes at each process. To find the bottleneck for each product, look for the procedure that takes the longest. As a result, task 4, which takes 14 minutes to process product A, is the bottleneck for product A, while task 6 is the bottleneck for product B.

To figure out the hourly capacity in products produced per hour, one must first assume that there will be no backlog, queue or waiting within the system. Product A's bottleneck is task 4, with a process length of 14 minutes, enabling the system to generate four products in an hour. Similarly, the capability to produce product B is three products per hour. If 60% of customers choose product A and 40% prefer product B, the company's average capacity is $(0.60 \times 4) + (0.40 \times 3) = 3.6$ product/hour with product units in terms of both A and B.

Although one can assume that there is no waiting while approaching to model an operations planning problem, queues often arise at bottlenecks, and all processes whose processing time is greater than the preceding processes. In this scenario, product A would be waiting before tasks 1, 2, 3, and 4 since the tasks preceding them have a greater rate of production as they process product A in a shorter length of time. Note the assumption that demand for product A is endless.

Similarly, product B would queue ahead of jobs 1 and 2 for the same reason. Naturally, both items may form queues ahead of task 1 since the arrival rate of demand may exceed the process rate of task 1.

Example 2

A company offers four types of products: A, B, C, and D. Allow these items to be processed in batches at five distinct workstations or operating zones. Product A is handled sequentially at tasks V (25), Y (5), and X (5); product B visits tasks Y (5) and X (15); product C travels via W (1), Z (1), X (1), and Y (1); and product D travels through W (10), Z (5), and Y (1). The numbers included in parenthesis represent the length of each product in minutes at each operation, and the demand for these items is 60, 80, 80, and 100 units, respectively. Table 2 displays the resultant workload for each activity by multiplying task time by product demand. This system's bottleneck is Task X, which has the greatest workload.

Table 2. Task loads and bottleneck.

	V	W	X	Y	Z
A	1500	0	300	300	0
B	0	0	1200	400	0
C	0	80	80	80	80
D	0	1000	0	100	500
Total	1500	1080	1580	880	580

Example 3

Jacobs and Chase (2018) present a case for explaining how to analyze the financial cost of manufacturing processes. A component is created by combining two parts: The first part, whose raw material costs 5 cents, is made in the shop using 5 machines, while the other part is brought in by an external source for 15 cents per part. Each machine that produces 20 pieces per hour is run by a person that receives 15 cents for each part they process in a typical shift of 9 hours per day, 5 days per week. This task has the following capacity: 5 machines × 20 parts/hour × machine × 9 hours/day × 5 days/week = 4500 parts/week.

During a typical shift, the firm employs 10 workers to assemble the first element with the second. This assembly line produces 100 components each hour. Each worker in the line receives 10 cents per component from the company. The following is the weekly capacity of the assembly process: 100 components per hour × 9 hours per day × 5 days per week = 4500 components each week.

A manufacturing process analysis might look for imbalances in process capacities. In this example, the capacity of the production and assembly processes are the same, 4500 units/week, showing that both jobs are balanced. If 7 machines were employed instead of 5, the manufacturing operation's capacity would have grown to 6300 parts/week (7 machines × 20 parts/hour × machine × 9 hours/day × 5 days/week). However, if the assembly capacity stays constant, the capacity of the whole operation will remain constant as well, since the operation's capacity cannot exceed the capacity of the slowest job, or the bottleneck. If another assembly shift were added to this situation, the assembly capacity would have grown to (100 components/hour × 18 hours/day × 5 days/week), making the manufacturing task the new bottleneck.

The production planner's primary responsibility is to calculate manufacturing expenses. Assume the business spends 1 cent per component for energy, $50 per week for rent, $500 per week for management, and lastly $25 per week for overall depreciation. Table 3 shows the average unit cost analysis for two scenarios in which the firm produces 4500 components per week in the first scenario and doubles output in the second.

According to economies of scale, the average unit cost falls as production volume grows. However, the opposite may be true in certain situations, and the diseconomy of scale refers to a situation in which an increase in production volume may result in an increase in unit cost.

Table 3. Unit cost comparison.

	Quantity	Price	Cost	Quantity	Price	Cost
Raw material	4500	0.05	225	9000	0.05	450
Purchased parts	4500	0.15	675	9000	0.15	1350
Molding labour	4500	0.15	675	9000	0.15	1350
Assembly labour	4500	0.1	450	9000	0.1	900
Management	1	500	500	1	500	500
Rent	1	50	50	1	50	50
Electricity	4500	0.01	45	9000	0.01	90
Depreciation	1	25	25	1	25	25
		Total cost	2645		Total cost	4715
		Unit cost	0.59		Unit cost	0.52

28

Companies often manufacture more than one kind of product, and in most cases, the production must be adjusted so that it is following the standards of the product or service being produced. The amount of time necessary to prepare an operation to produce a different good or service is referred to as the setup time. It is the time spent in preparing an operation to start or changing the way that it functions, and it is sometimes referred to as changeover time. Time has a monetary value in business, which is known as the interest rate. In a similar vein, the setup cost and, by extension, the production cost will increase proportionally with the length of time needed for the setup. In the preceding illustrative situations, setup times were not considered. The following example shows a fundamental knowledge of setup time.

Example 4

Manufacturing process analysis can help the planners to identify the cost of time, which can also be used as a template for service operations. Jacobs and Chase (2018) present another case centered around the time cost where a company makes two kinds of parts: part 1 and part 2. A machine is used to process parts, however only one sort of part may be made at a time. The machine's production rate is 25 seconds for part 1 and 20 seconds for part 2. The production planner has decided that 50 units of each item will be worked at the machine. In other words, the batch size for each part is 50 units. The machine costs $400 per hour to run and moving from one element to another involves changeover time. Preparing the machine for parts 1 and 2 takes 60 seconds and 70 seconds, respectively. The production planner should compute the total time needed to make a batch of each item to figure out the unit cost of the product that has both elements.

To calculate the average unit cost, one first needs to calculate the overall cost of the operation for the complete set of components produced within a certain time. The overall operating cost in this example refers to the machine run time, and the amount of parts produced throughout a period is determined by the following equation, which includes the setup time for each component and the run time required to output the batch size: $60 + 70 + 25(50) + 20(50) = 2380$ seconds/50 units. This production rate is converted to an hourly (3600 seconds) output rate using the following ratio: $2380/50 = 3600/x$. The machine's output is 75 units/hour after solving for x. As a result, the unit cost is $400/75 = $5.3/unit.

A critical choice for the production planner here is finding the ideal batch size that ensures the lowest unit cost. There are extensive optimization and simulation approaches for solving this issue, but by just giving an arbitrary batch size, one can understand the mechanics of this choice. To investigate the change in unit cost, set the batch size to 75 units: $60 + 70 + 25(75) + 20(75) = 3505$ seconds/75 units. As a result, the machine's production rate is 51 units per hour, and the unit cost is around $7.8. This manufacturing cost is much higher than in the situation when the batch size is 50.

A break-even analysis is a popular production approach in which the term break-even refers to the point at which cost equals income. Many examples of this fundamental analysis can be found in operations management and research literature (Albright and Winston, 2017; Anderson et al., 2016; Cachon and Terwiesch, 2012; Dowling, 2009; Evans, 2017; Heizer et al., 2017; Kobu, 1996; Krajewski et al., 2016; Nahmias and Olsen, 2015; Taylor and Russell, 2011; Winston and Albright, 2019).

Example 5

This example presents a simplified version of the case in (Jacobs and Chase, 2018) which analyzes a company that intends to enhance its manufacturing line by introducing two new robots. These robots should perform the work of four humans, each earning $30,000 per year. A worker will run the robots and earn $60,000 a year. The new equipment costs $300,000, but the contemporary design will save $0.2 for each item produced. Leaving out the value for money over time, one can find the company's break-even point in terms of the number of items it makes at the end of the second year.

Current operations cost $30,000/worker × 4 workers × 2 years = $240,000. For the following two years, the cost of the new operational design is $60,000/worker × 2 + $300,000/robots − $0.2 × x, where x is the production volume of items. Equating these two alternative possibilities yields the output volume: $240,000 = $420,000 − $0.2 × x.

To calculate the production volume of items or break-even threshold, divide $180,000 by $0.2 to get 900,000 items. This leads one to the conclusion that the company will only be able to compensate for the new operational design by manufacturing 900,000 items during the period of the investment.

The ideal manufacturing environment is one in which a corporation produces massive quantities of an unstandardized item, often known as mass customization. An inefficient process, on the other hand, relates to creating highly standardized things in small numbers. Figuring out the production environment is a precondition for stylistic layout applications, and the next chapter describes layout studies for production settings.

References

Albright, S. and Winston, W. (2017). *Business Analytics: Data Analysis and Decision Making* (6th Edn.). Boston, Cengage Learning.

Anderson, D., Sweeney, D., Williams, T., Camm, J., Cochran, J., Fry, M. and Ohlmann, J. (2016). *Quantitative Methods for Business (Book Only)* (13th Edn.).

Cachon, G. and Terwiesch, C. (2012). *Matching Supply with Demand: An Introduction to Operations Management.* McGraw-Hill/Irwin.

Chase, R., Robert, F. and Nicholas, J. (2006). *Operations Management for Competitive Advantage* (11th Edn.). McGraw-Hill/Irwin.

Dowling, E.T. (2009). *Schaum's Outline of Mathematical Methods for Business and Economics.* McGraw Hill Professional.

Evans, J.R. (2017). *Business Analytics.* Pearson.

Goldratt, E. (1980). Optimized production timetable: A revolutionary program for industry. *APICS 23rd Annual Conference Proceedings,* 14–17.

Heizer, J., Render, B. and Munson, C. (2017). *Operations Management: Sustainability and Supply Chain Management* (12th Edn.). Pearson Education Limited.

Jacobs, F.R. and Chase, R.B. (2018). *Operations and Supply Chain Management* (15th Edn.). McGraw-Hill/Irwin.

Kasapoğlu, Ö.A. and Tayalı, H.A. (2012). Transformation of job shop to flow shop in an era of global crises. *Proceedings of the 10th International Logistics & Supply Chain Congress,* 112–116. https://papers.ssrn.com/sol3/papers.cfm?abstract_id=2919847.

Kobu, B. (1996). *Production Management.* Istanbul University, School of Business.

Krajewski, L., Malhotra, M. and Ritzman, L. (2016). *Operations Management: Processes and Supply Chains* (11th Edn.). Pearson.

Nahmias, S. and Olsen, T. (2015). *Production and Operations Analysis* (7th Edn.). Waveland Press.

Tayalı. (2021a). Manufacturing scheduling strategy for digital enterprise transformation. pp. 104–119. *In*: K. Sandhu (ed.). *Emerging Challenges, Solutions, and Best Practices for Digital Enterprise Transformation.* IGI Global. https://www.igi-global.com/chapter/manufacturing-scheduling-strategy-for-digital-enterprise-transformation/275703.

Tayalı, H.A. (2016). A literature review on production scheduling with the drum-buffer-rope technique. *16th International Symposium for Production Research,* 1085–1090. https://papers.ssrn.com/sol3/papers.cfm?abstract_id=2919846.

Tayalı, H.A. (2021b). A novel web-based decision support system for aggregate production planning problem. pp. 135–153. *In*: F. Saruchera (ed.). *Advanced Perspectives on Global Industry Transitions and Business Opportunities.* IGI Global. https://www.igi-global.com/chapter/a-novel-web-based-decision-support-system-for-aggregate-production-planning-problem/274913.

Taylor, B.W. and Russell, R.S. (2011). *Operations Management: Creating Value along the Supply Chain*. John Wiley & Sons.

Turban, E., Sharda, R. and Delen, D. (2014). *Business Intelligence and Analytics: Systems for Decision Support* (10th Edn.). Pearson.

Winston, W.L. and Albright, S.C. (2019). *Practical Management Science* (6th Edn.). Cengage.

CHAPTER 4
Facility Layout and Line Balancing

The physical arrangement of manufacturing operations in relation to each other is called the problem of facility layout. It is a mathematically difficult problem, thus, a challenging task for the production planner. However, the problem itself serves as an introduction to the abstract thinking ability required by a planner: The ability to model and represent operational planning challenges.

Even though the layout of a manufacturing environment is typically figured out by the pattern of workflow, there are four different layout types or strategies that can be used to place manufacturing processes and operations in relation to each other: the manufacturing cell type, assembly line type, project type, and work center type. It should not be surprising for a production planner to see a facility harboring more than a single type of layout. Eventually, because it is usually more efficient to design the facility by the path that a product follows within the environment, most manufacturing facilities use a combination of multiple layout types. Thus, a part of a facility might be organized as a work center, while another place as a manufacturing area dedicated to processing a part, and another area designed to follow the working guidelines of an assembly line, for instance, a couple of designated machines connected with a conveyor belt.

One of the primary layout designs is referred to as a project layout, also known as a fixed position layout. When speaking of a work center layout, the production literature often makes use of phrases such as "job shop" and "functional layout". The layout design of a manufacturing cell is a hybrid layout since it incorporates the design ideas of both a work center and an assembly line into its architecture. In other words, a manufacturing cell is designed to perform a set of processes just like a work center, and it works with a product-oriented approach at the same time, instead of the process-oriented nature of a work center. As a pure product-oriented layout the assembly line design is the last fundamental strategy discussed in this chapter.

Physically arranging industrial activities in relation to one another is a difficult assignment for the production planner. A facility may have multiple layout types, and manufacturing facilities might be working with more than one layout type since it is more practical to structure the facility according to the route that a product follows. For example, one region of the facility may be appointed as a work center, another as a production area devoted to processing a part, and still another as an assembly line. Offices, retail outlets, and warehouses are the core layout methods for service businesses. Detail planning is needed for both production and service activities to perfect their layout in line with business goals. This chapter provides the reader with a mathematical technique for layout design in assembly lines and introduces the various layout types for manufacturing processes. Although the architecture of a production environment is primarily figured out by the workflow pattern, there are four fundamental layout types or strategies for manufacturing processes (Jacobs and Chase, 2018; Kasapoğlu and Tayalı, 2012; Tayalı, 2021):

- Project
- Work center
- Assembly line
- Manufacturing cell

Project layout, also called a fixed position layout, is one of the basic layout designs. It is called a fixed position layout because the product stays fixed at a location. The equipment and materials are positioned within or around the facility. Construction sites are the simplest example for this type of layout.

Another fundamental layout concept employed in facilities where process-oriented functions are conducted is the work center. Because a work center does a certain sort of work, it is sometimes referred to as a department. When referring to a work center layout, words such as a job shop or a functional layout are often used in manufacturing literature. This arrangement is proper for make-to-order or engineer-to-order manufacturing environments in which the company normally develops goods directly from raw materials and components and the material flow changes from product to product. Similar machinery, equipment, or functions are positioned next to each other in this layout style, and a certain sort of activity is conducted in its own zone. The goal is to reduce the cost of material handling while the processed goods continue their route. Hospitals and woodworking businesses are common examples of work center layout design. When selecting the ideal layout design for a work center, companies should have updated data on the flow of goods and routes connected to materials handling. Furthermore, in certain circumstances, the work center layout issue is analogous to the traveling salesperson problem, a classic problem, which is computationally demanding and needs complex mathematical approaches to solve (Winston and Albright, 2019).

The assembly line, a product-oriented layout technique, is the last fundamental layout concept examined in this chapter. In this arrangement, an item travels a manufacturing path and visits a set of work centers. There may also be a material handling device, such as a belt or roller conveyor, but the basic premise of an assembly line layout design is that all workstations must keep up with the predetermined output rate to function synchronously and in balance. After ensuring that permissible job time is equal at each workstation and reaching balance, the layout of the workstations, machinery, equipment, or processes is structured. Assembly line layouts are, of course, proper for facilities that produce using make-to-stock or assemble-to-order methods and use a variety of flow types of manufacturing processes. The difficulty of balancing an assembly line is discussed further in the examples at the end of the chapter.

Manufacturing cell layout design is a hybrid layout method that incorporates both work center and assembly line design ideas. To produce items with comparable standards, dissimilar machinery and equipment are arranged inside a specialized space called a manufacturing cell, or a work cell. A cell, like a work center, is intended to complete a series of tasks and works with a product-oriented approach, like an assembly line. As a result, manufacturing cells have the advantages of minimal material handling expenses and work-in-process inventories.

Line balancing

A line is made up of workstations. Parts and products continue toward completion by following a predefined route across these workstations. Each job in a balanced line belongs to a workstation; that is, each workstation includes tasks where humans or machines work on the components or the products. Balancing a line is synchronizing how these workstations function, and a production planner requires an analytical approach to address this line-balancing challenge. Following this process of synchronization, and balancing the line, the architecture of the facility that runs a line is planned to reduce the expenses (such as material handling and transfer) that may be incurred due to the distances between these workstations.

A set of mathematical computations constitutes an algorithm. Following a series of calculations to fulfil a mathematical task is like following a recipe to create a food. The goal of the line-balancing issue is to distribute jobs to workstations such that at the completion of a cycle period, each station outputs its task. Take note that cycle time is the same as lead time, which is the period between the start and finish of a project. However, in the context of an assembly line, it is defined as the interval between units finished at each station. In other words, it is the largest amount of time that each station may devote to working on a part or a product. It is figured out algorithmically by dividing available manufacturing time by demand. on the other hand, finding the cycle time might be the simplest calculation since it is the precedence connections between the jobs that complicate the issue of line balancing. As a result, the first stage in line balancing is to map the sequential or precedence links between the line's jobs. These precedence relationships develop naturally in the manufacturing function because each part or product has a distinct product design and so a separate processing path. Precedence relationships define the sequence of processing for an item, or the order in which operations are done on the object. Again, there are many examples of balancing an assembly line in operations management and research literature.

Example 6

Following the precedence relationships and task timings presented in Table 4, a company may produce 475 balls each day in the allotted production time of 450 minutes (Jacobs and Chase, 2018).

Table 4. Data for ball line-balancing problem.

Task	Duration seconds	Preceding tasks
A	44	-
B	10	A
C	8	B
D	49	-
E	14	D
F	11	C
G	11	C
H	11	E
I	11	E
J	7	F, G, H, I
K	8	J

When looking at the data, it might be difficult to keep track of the complete set of precedence connections. A task is said to have an immediate predecessor if it is a task that must be finished before another task, so that the next one can start. For instance, one may figure out that task E is an immediate predecessor of task I, and that task D is an immediate predecessor of task E. Both relationships can be found in the task structure. Nevertheless, one cannot simultaneously see the interconnectedness of jobs I and D in the same way. As a result, using a precedence diagram helps the planner, before production starts, to see, and understand the order of duties, and the position of workstations in relation to one another, and so the complete layout, the structure, and the final design of the assembly line.

The construction of a precedence diagram is the first stage of the algorithm evaluation process for line balancing. Image 1 outlines the fundamental precedence relationships seen in a production environment. One can see, on the left side of the picture, that activity A is a direct predecessor of activity B, and similarly, task B is a direct predecessor of activity C. Since activities A and B come before job C in the middle, it is impossible to go on to activity C until A and B have been completed. On the right, in a manner analogous to what was just described, one can see that A is a direct predecessor of both B and C, and that neither B nor C can begin until A completes its task first. The numbers in Image 2 stand for the length of each job in the precedence diagram for the ball example.

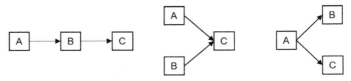

Image 1. Precedence relationships.

The balance of an assembly line has a dual purpose in the overall process. First, to organize the stations in such a way as to produce a plan that reduces the distances that separate these groups, and second, to make sure that the desired output rate is met within the allotted amount of production time. It should be

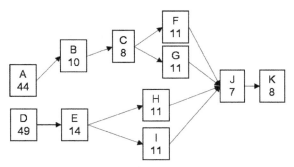

Image 2. Precedence diagram for the ball example.

emphasized once again that these goals need to be in line with the precedent relationships that exist between the activities. If the workstations were not grouped together, the processes would have been scheduled to conclude at the earliest possible moment, which would be the 184th second:

Sum of task times = 44 + 10 + 8 + 49 + 14 + 11 + 11 + 11 + 11 + 7 + 8 = 184 *seconds/ball.*

If this is the case, then the product will make its way through task A, then task D, then either task B or task E, and so forth until it reaches task K. The concept behind line balancing is to work in parallel at the stations that are going to be created. This is done based on the cycle time, which is estimated based on the parameters of available production time and needed demand.

$$Cycle\ time = \frac{Available\ production\ time}{Required\ demand} = \frac{450\ minutes * 60\ seconds}{475\ balls}$$
$$= 56.8 \cong 57\ seconds/ball$$

According to the predicted cycle time, each station that is going to be set up has 56.8 seconds, or 57 seconds to simplify things, accessible to them to do the duties that are going to be given to them. To put it another way, the line will produce one unit every 57 seconds to fulfill the necessary demand over the whole of the manufacturing day. Instead of processing a ball in a sequential manner on the line and spending a minimum of 184 seconds doing so, the production may be parallelized such that a unit comes off the line every 57th second. This reduces the time needed to complete the operation from 184 seconds to 57 seconds.

The next thing that must be done to balance a line is to figure out the number of parallel stations required by the line. The fundamental idea is to maximize efficiency by reducing the number of stations needed to complete a given level of production so that no other labor is needed. Keeping in mind Occam's razor, one should avoid adding more entities than are necessary to explain anything (James et al., 2021). The theoretical minimum number of stations is the least number of stations that is theoretically achievable and is figured out by dividing the entire time necessary to manufacture one unit (a ball, in this case) by the cycle time:

$$N_t = \frac{Total\ task\ time}{Cycle\ time} = \frac{184\ seconds/ball}{57\ seconds/ball} = 3.2 \cong 4$$

It is important to note that the resultant number does not have any units; rather, it just shows the number of stations that need to be formed to distribute the jobs among these stations. Since it is inconvenient to first build a fraction of a station and then distribute a fraction of tasks into a fractional station, the number that is generated is always rounded up to the next integer.

The line-balancing issue has, up to this point, reduced the total number of stations while adhering to the cycle time limit; nevertheless, the precedence connections have not played a substantial part in the process. The next thing that must be done is to divide up the work across the four different stations that are going to be set up. There are three primary considerations involved in doing so, the first of which is to adhere to the cycle time, also known as the restricted amount of time, allowed at each station, which is 57 seconds: It is necessary to delegate duties until the station can no longer accommodate any more assignments owing to capacity constraints. The second argument is that jobs cannot be given randomly to stations owing to precedence relationships; hence, there may be more than a single solution to the problem, depending on the number of available different sequences. When finding the order in which tasks are to be assigned, one may use the process of trial and error, but there are other algorithmic methods and heuristic decision rules that can be used. These can be used to decide how to distribute the next task to a workstation that is going to be formed. To figure out which stations manage which tasks, one needs to organize the stations according to the amount of time it takes them to do their responsibilities or the number of tasks that come after them, all the while keeping in mind the precedence connections. In situations when there is a tie between the jobs decided by the main assignment rule, such as the scenario with the ball, it is possible that a secondary assignment rule might also be needed. For example, the first thing that must be done is to find which of the tasks labeled F, G, H, or I is going to be delegated to the correct station. Therefore, if there is still a tie after the application of the main assignment rule, a secondary assignment rule may be used to break the ties. Again, the production planner decides on the secondary assignment rule, if not a machine.

In short, selecting a primary and a secondary assignment rule is the next step for balancing the line, after one has determined the theoretical minimum number of stations needed. The primary rule can be an arbitrary decision rule, i.e., shortest task time, and

Table 5. Primary and secondary assignment rules.

Task	Number of following tasks	Longest task time
A	6	
B or D	5	D
C or E	4	E
F, G, H, or I	2	F, G, H, or I
J	1	
K	0	

the secondary rule can again be an arbitrary decision rule, i.e., the task with the fewest number of follower tasks. Table 5 shows the application of primary and secondary assignment rules for the ball example, where descending number of followings tasks are used as a primary assignment rule, and the longest task time as the secondary assignment rule. One can easily trace and record a similar table using the precedence diagram. For instance, task A has 6 following tasks, namely, B, C, F, G, J, and K.

Tasks are then distributed, beginning with, and progressing through station 1, until either the total amount of time spent on given tasks equals the cycle time, or no more tasks are possible to assign owing to the constraints imposed by the cycle time limit or the precedence relationship. It is important to keep cycling through the assignment process until all jobs have been delegated to their proper locations. The complete ball line balancing solution is shown in Table 6 (note the tasks selected arbitrarily where both assignment rules lead to a tie).

Image 3 depicts a conceptual layout design for the assembly line example consisting of 4 stations represented by circles. If facilities, machinery, equipment, worker abilities, and other associated elements fit the needs of the job, a production planner may consider arranging tasks inside designated stations or

Table 6. An alternative solution for the ball example.

Station	Task	Task time	Remaining unassigned time	Feasible remaining tasks	Tasks with most followers	Tasks with longest task time
1	A	44	13	B		
	B	10	3 (idle)	None		
2	D	49	8 (idle)	None		
3	E	14	43	C		
	C	8	35	F, G, H, I	F, G, H, I	F, G, H, I
	F	11	24	G, H, I	G, H, I	G, H, I
	G	11	13	H, I	H, I	H, I
	H	11	2 (idle)	None		
4	I	11	46	J		
	J	7	39	K		
	K	8	31 (idle)	None		

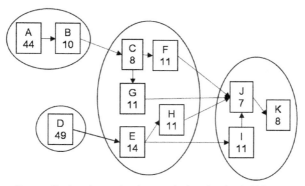

Image 3. An alternative layout design for the ball line.

grouping them near together to reduce the distance between each interdependent activity and, therefore, costs.

Table 6 shows that the stations have downtime. For instance, station 1 has 3 seconds of idle time since no other job may be added to it owing to cycle time or precedence relationship constraints. The efficiency of the line decreases as the idle team increases, and the efficiency of a line is calculated by the following formula:

$$Efficiency = \frac{Total\ task\ time}{N_\alpha * Cycle\ time} = \frac{184\ seconds/ball}{4*57\ seconds/ball} = \%\ 80.7$$

N_t is the theoretical minimum number of stations and N_α refers to the actual number of stations formed within the assembly line. The actual number of stations might be bigger than the theoretical minimum number of stations. It is important to note that there is no predefined rule that guarantees the most efficient line balance. Despite this, there are software packages available for line balancing problems that should lead a production planner to the best efficiency and thus layout design.

Example 7

Say a corporation works 500 productive minutes each day to produce a toy for which the manufacturing plan calls for 50 units. Table 7 shows the precedence connections for producing the

Table 7. Precedence relationship for the toy example.

Task	Duration (minutes)	Preceding tasks
A	9	-
B	10	A
C	4	B
D	3	B
E	10	A
F	2	C, D
G	6	F
H	10	E
I	2	G, H

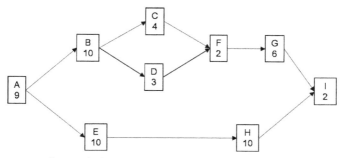

Image 4. Precedence diagram for the toy example.

toy, and Image 4 displays the precedence diagram (Heizer et al., 2017).

Calculating the cycle time and theoretical minimum number of stations that are going to be formed:

$$C = \frac{Available\ production\ time}{Required\ demand} = \frac{500\ minutes}{50\ toys} = 10\ minutes/toy$$

$$N_t = \frac{Total\ task\ time}{Cycle\ time} = \frac{56\ minutes/toy}{10\ minutes/toy} = 5.6 \cong 6$$

There are two main requirements that must be followed to evaluate the work as a candidate for assignment to a station. First, all its predecessors must have been given to a station, and cycle time cannot be surpassed. The primary assignment heuristic in this case is the lowest processing time, while the secondary is the fewest followers. Table 8 illustrates an alternate solution for the toy example and a different reporting technique for work assignment.

In Table 8, the abbreviation "N/A" stands for "not applicable", writing down that the candidate task is not subject to assignment heuristics. Image 5 depicts an alternative layout plan that places jobs in the same stations next to each other to reduce distance-related expenses inside the production facility.

Table 8. An alternative solution for the toy example.

Station	Candidate tasks	Shortest processing time	Fewest follower	Assigned task	Task time	Remaining unassigned time
1	A	N/A	N/A	A	9	1 (idle)
2	B, E	B, E	E	E	10	0
3	B, H	B, H	H	H	10	0
4	B	N/A	N/A	B	10	0
5	C, D	D	N/A	D	3	7
	C	N/A	N/A	C	4	3
	F	N/A	N/A	F	2	1 (idle)
6	G	N/A	N/A	G	6	4
	I	N/A	N/A	I	2	2 (idle)

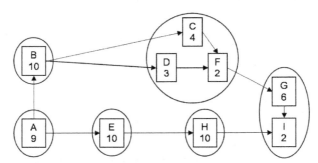

Image 5. An alternative layout design for the toy line.

Finally, the line's efficiency:

$$Efficiency = \frac{Total\ task\ time}{N_\alpha * Cycle\ time} = \frac{56\ minutes/toy}{6*10\ minutes/toy} = \%\ 93.3$$

The production planner is introduced to the abstract thinking necessary for modeling operations planning issues when confronted with the mathematical computations involved in designing a facility. Note that the required level of inventory to sustain production is neglected in the above examples. The techniques described in this chapter, such as precedence diagrams and algorithmic design, will be essential in the next chapter, which covers scheduling projects and operations.

References

Heizer, J., Render, B. and Munson, C. (2017). *Operations Management: Sustainability and Supply Chain Management* (12th Edn.). Pearson Education Limited.

Jacobs, F.R. and Chase, R.B. (2018). *Operations and Supply Chain Management* (15th Edn.). McGraw-Hill/Irwin.

James, G., Witten, D., Hastie, T. and Tibshirani, R. (2021). *An Introduction to Statistical Learning*. Springer, US. https://doi.org/10.1007/978-1-0716-1418-1.

Kasapoğlu, Ö.A. and Tayalı, H.A. (2012). Transformation of job shop to flow shop in an era of global crises. *Proceedings of the 10th International Logistics & Supply Chain Congress*, 112–116. https://papers.ssrn.com/sol3/papers.cfm?abstract_id=2919847.

Tayalı. (2021). Manufacturing scheduling strategy for digital enterprise transformation. pp. 104–119. *In*: K. Sandhu (ed.). *Emerging Challenges, Solutions, and Best Practices for Digital Enterprise Transformation*. IGI Global. https://www.igi-global.com/chapter/manufacturing-scheduling-strategy-for-digital-enterprise-transformation/275703.

Winston, W.L. and Albright, S.C. (2019). *Practical Management Science* (6th Edn.). Cengage.

CHAPTER 5
Project Scheduling

A project is a collection of interdependent activities or bits of work that are constrained and reliant on a set of predefined production factors. The term project is composed of the suffix '-ject', which means to throw, and the prefix 'pro', which implies before or earlier than. The goal of planning a project and its tasks is to estimate the completion date for monitoring and regulating purposes.

A systematic approach to a project could aid a production planner in efficiently managing a project. The essential tool an operations planner can use to track and manage a project in a stylish manner is a Gantt chart. This chapter supplies numerical examples of the critical path technique for project planning and scheduling.

The Critical path method finds scheduling information for each activity within a project for improved production process and resource planning, monitoring, and control. To schedule the project's tasks, the method figures out when each task begins and ends. To estimate the project's completion time and the schedules for each work, a production planner must define the order of operations, preferably using a precedence diagram that depicts the relationships between the tasks. The critical network route is then found by analyzing each task sequence from the beginning to the end of the project.

The prefix 'pro' implies before or ahead of, and the suffix '-ject', which means to throw, creates the term project. To summarize, a project is the process of expecting and preparing for future events. Typically, projects begin and conclude on their designated dates, while unexpected events may and do occur. Formally, a project is a collection of interconnected tasks or pieces of work that consume and are constrained by production factors. The completion of all tasks concludes the project and generates a result.

If a production planner takes a methodical approach to a project, it may be easier for them to manage the projects. As such, a Gantt chart, first developed by Henry L. Gantt at the turn of the 20th century and initially used to shipbuilding projects, is the fundamental tool that a planner may use throughout a project to watch and govern it in a stylish manner (Wilson, 2003). Gantt charts are still often used by production planners and software packages for the purposes of planning and monitoring projects, as well as scheduling and assigning resources within the shop floor. Again, the most important thing is to make use of the data collected from the shop floor while planning operations. Image 6 is a sample of a Gantt chart that illustrates the relationship between various jobs that are part of a production schedule.

Even though the tasks on the Gantt progress chart in Image 6 seem to be moving along without any problems, the reality is that things do not go according to plan, usually. There is a possibility that multiple jobs may overlap with one another or fall behind time. There are sophisticated tools for Gantt charting that display

Image 6. A Gantt chart example.

not just when a work begins and when it concludes, but also the actual progress that is being made on the task, as well as any delays and periods when it is not productive.

Critical path method

After the introduction of the Gantt chart, two well-known models for network planning have been developed:

- Critical path method (CPM)
- Program evaluation and review technique (PERT)

Both models were developed in the middle of the 20th century. CPM was developed by DuPont, a company that owns chemical processing plants, to schedule maintenance work at these plants, while PERT was developed through a military effort (Kelley and Walkerf, 1959). Both CPM and PERT aim to estimate uncertain project completion timelines with a degree of precision. Compared to CPM, PERT needs more of a statistical perspective, particularly about the foundations of probability distributions of normal and beta, which are left outside the scope of this book.

CPM is a similar algorithm to the assembly line-balancing algorithm, but it serves a different purpose. The aim of the assembly line balancing is to boost productivity and efficiency, and to recommend an efficient layout design for the line, while the aim of CPM is to schedule a project with tasks with predefined start and end times. In CPM, one works again with precedence diagrams with task times, but what distinguishes it from assembly line balancing is that CPM serves for task scheduling. In short, CPM figures out when tasks will begin and end, and it also has a property called the critical path, which is the longest path within the set of task routes.

To begin using CPM, one needs to define the project by creating its outline and specifying the set of interrelated tasks of the production process. A precedence diagram is used with few modifications, but the production planner must ensure that the preceding and following tasks, in other words, immediate predecessor and following tasks, are properly defined at the outset. The precedence diagram is depicted as a network connecting all tasks, with time estimates assigned to each task. The critical path is then defined, which is the sequence of network tasks that takes the longest to complete. The importance of the critical path stems from the fact that if any task on this path is delayed, the entire project will be delayed. It should be noted that a network may have multiple critical paths.

As previously told, the goal of the CPM is to figure out the scheduling information for each task of the project to better plan, monitor, and control the production process and resources. In the operations management and research literature, there are numerous examples of project scheduling utilizing CPM and other approaches such as PERT.

Example 8

Consider a project with four tasks and their duration as given in Table 9.

Table 9. Data for project scheduling example.

Task	Duration (weeks)	Preceding tasks
A	2	-
B	1	A
C	3	A
D	4	B, C

A production planner must decide the sequence of activities, preferably using a precedence diagram that reflects the relationships between the tasks, to estimate the project's completion time and schedules for each task in the project. Image 7 depicts the precedence diagram for the project's network, which consists of four tasks, based on the data presented in Table 9.

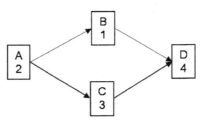

Image 7. Precedence diagram for project scheduling example.

The network's critical path is figured out by examining each sequence of tasks that flows from the beginning to the end of the project. In this example, there are two alternative task sequences, or paths, as described: A-B-D and A-C-D, but only the path of A-C-D is the critical path because it takes 9 weeks to complete this path, whereas the task times for the other path, namely, A-B-D, add up to 7 weeks. If there is a delay along the critical path, or if any of the tasks of A, C, or D is not completed on time, there will be a delay in the scheduled project completion time, which will be calculated in the next step.

The CPM calculates when each task begins and ends to schedule the project's tasks. The method finds four separate times for each task in the project for this purpose:

- Early start (ES): The earliest time that a task can start.
- Early finish (EF): The earliest time that a task can end.
- Late start (LS): The latest time that a task can start.
- Late finish (LF): The latest time that a task can end.

Note that ES cannot be calculated for a task until all its predecessors have been completed. Image 8 shows a way for representing a task as a network node with all pertinent information around it. Note that the task's name and duration are defined in the center of this node.

ES ⌒ EF
A
2
LS ⌣ LF

Image 8. A node in the project network.

CPM works in two steps. The first step, or forward pass, figures out ES and EF for each task, while the second set of calculations, or backward pass, calculates LS and LF. Because of the project's structure or allocation of production resources, a task may have slack time or leeway between its starting and ending times. To put it another way, the slack time of a task is the time between the late and early times of the task. If a task has slack time, a production planner can delay that task for its slack time without causing a delay in the completion of the entire project. The critical path has no slack time, as expected. In fact, another definition of the critical path is a path on which all tasks have zero slack time.

Starting with the first task, the ES for task A is 0. Since it must be processed for a maximum of 2 weeks, one needs to add that amount with that of ES to get 2 weeks as the EF time. In other words, task A needs to start at week 0, and to finish at week 2, at its earliest. The LS and LF times are calculated during the backward pass, after all earliest times of tasks are calculated in this manner. Image 9 shows the project's network diagram following the first pass.

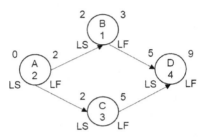

Image 9. Project's network diagram, forward pass.

Constructing the early schedule proceeds with task B, after task A. The EF(A) = 2, so ES(B) = 2. Similarly, EF(A) = ES(C) = 2. Both B and C have task A as their immediate predecessor. However, there is a different case for task D, where it has two immediate predecessors, namely B, and C. Therefore, ES(D) = EF(C) since C finishes later than B, and D can only start when B and C both finish.

While starting the backward pass to calculate the late times, CPM accepts the last task's EF as the task's LF. Thus, EF(D) = 9 and LF(D) = 9 according to the network diagram in Image 9. Since D has four weeks of processing time, one needs to subtract those four weeks from the task's latest finish time, LF(D) = 9, to get LS(D) = 5. As a result, task D should start by the 5th week at the latest, or the project may be delayed. The same logic of the forward pass applies in the reverse direction. For instance, A has two immediate predecessors while considering the network backwards, and LF(A) = LS(C) because C, by the latest, should start at an earlier time than B. This forces A to start in week 2 at the latest. All that leads to a slack time of two weeks on task B since LS(B) – ES(B) = LF(B) – EF(B) = 2. Therefore, the other path of A-C-D is the critical path because it does not have a task with a slack time. Image 10 presents the final network diagram with all times.

Note that manually generating the reverse schedule can be error prone. There are software packages that automatically report the findings for the user after the data has been input. The results of this example are presented in Table 10, where task B has slack time between its starting and finishing times and is therefore not on the critical path.

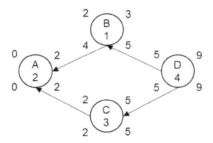

Image 10. Project's network diagram, backward pass.

Table 10. Final schedule of the project.

Task	ES	EF	LS	LF	LS-ES	LF-EF	CP
A	0	2	0	2	0	0	Yes
B	2	3	4	5	2	2	No
C	2	5	2	5	0	0	Yes
D	5	9	5	9	0	0	Yes

Jacobs and Chase (2018) suggest that a project may be challenging to manage if it contains multiple critical routes. The next example contains a single critical path.

Example 9

Consider a project with seven tasks and their duration as given in Table 11.

Table 11. Project data.

Task	Duration (weeks)	Preceding tasks
A	20	-
B	4	A
C	6	A
D	1	B, C
E	4	C, D
F	7	C, D
G	1	E, F

The aim in scheduling a project and its tasks is to estimate its completion time for monitoring and controlling purposes. As previously mentioned, a precedence diagram should reflect the sequential relationships between the tasks of a project. Image 11 shows the precedence diagram, or the network of the project without a schedule. Images 12 and 13 show the project's early and late schedules as figured out by the CPM's forward and backward passes.

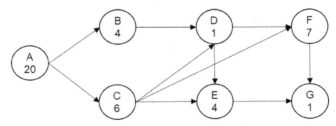

Image 11. Precedence diagram without a schedule.

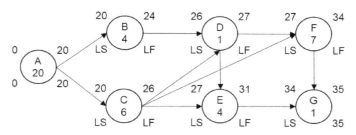

Image 12. Forward pass.

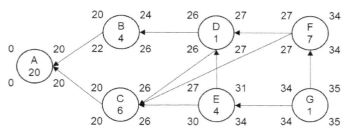

Image 13. Backward pass.

In this project, based on the precedence relationships, one can find 5 alternative paths from the start (task A) to the end (task G) with the following completion times:

- A-B-D-E-G: 20 + 4 + 1 + 4 + 1 = 30 weeks
- A-B-D-F-G: 20 + 4 + 1 + 7 + 1 = 33 weeks
- A-C-D-E-G: 20 + 6 + 1 + 4 + 1 = 32 weeks
- A-C-D-F-G: 20 + 6 + 1 + 7 + 1 = 35 weeks
- A-C-F-G: 20 + 6 + 7 + 1 = 34 weeks

There is no room for delay along the critical path A-C-D-F-G, as this path takes the longest time to complete compared to alternate routes. Table 12 confirms that this is the path with zero slack time and presents the project's detailed schedule in a tabular format.

Table 12. Final schedule of the project.

Task	ES	EF	LS	LF	LS-ES	LF-EF	CP
A	0	20	0	20	0	0	Yes
B	20	24	22	26	2	2	No
C	20	26	20	26	0	0	Yes
D	26	27	26	27	0	0	Yes
E	27	31	30	34	3	3	No
F	27	34	27	34	0	0	Yes
G	34	35	34	35	0	0	Yes

The concept of scheduling is central to operations planning. The following chapter examines the scheduling activities of a manufacturing job shop, where sequencing comes into play more.

References

Jacobs, F.R. and Chase, R.B. (2018). *Operations and Supply Chain Management* (15th Edn.). McGraw-Hill/Irwin.

Kelley, J.E. and Walkerf, M.R. (1959). Critical-path planning and scheduling. *Proceedings of the Eastern Joint Computer Conference*, pp. 160–173.

Wilson, J.M. (2003). Gantt charts: A centenary appreciation. *European Journal of Operational Research*, 149(2): 430–437. https://doi.org/10.1016/S0377-2217(02)00769-5.

CHAPTER 6
Work Center Scheduling

Most facilities around the world have a job shop structure, with output that is highly variable and produced in small quantities. Scheduling is a critical part of these facilities' operations. Meeting agreed-upon deadlines, for example, has a direct relationship with customer satisfaction. Thus, it presents a difficult problem for the operations planner, often known as the scheduler, to make the best scheduling selections.

It should be noted that orders and work centers on the shop floor can be used to gather the data needed for scheduling, making this another area where digital data is crucial. Scheduling a work center finds out not only when jobs begin and end at the work center, but also the sequence of jobs that are fed into that center: A scheduler uses priority rules, also known as heuristics, to order — or sequence — the jobs assigned to the work center.

Depending on the structure of the production environment, a scheduler might come across four distinct scenarios: (1)

Scheduling n jobs on a single workstation; (2) Scheduling n jobs on two different work centers; (3) Scheduling n jobs across m number of multiple work centers; (4) Scheduling n tasks across n number of work centers. The most basic type of operations scheduling problem is sequencing a series of jobs on a single work center, but the last two industrial settings demand sophisticated mathematical models to generate a correct schedule. The chapter concludes by delving into various scheduling assumptions, such as whether work centers can process the same jobs concurrently or not.

Scheduling is an essential part of operations planning and a demanding responsibility of the production planner or the scheduler. The scheduling process is very susceptible to outside influences, such as the absence of a worker or the malfunction of a piece of equipment. Due to the unpredictability of these external elements, a production planner or scheduler cannot always ensure that the schedule will be met, but by paying attention to how the goods or services are designed and delivered, they may enhance the associated operations scheduling activities. Among the key duties of the scheduler are watching the facility, monitoring feedback for unfinished work, and continually updating and rescheduling assignments.

From restaurants to hospitals, most facilities in the world have a job shop structure in which output is highly variable and produced in low amounts. Chase et al. (2006) define job shops with low production volumes and assume infinite production capacity while modelling a job shop scheduling problem.

A job shop's production environment is optimized for the make-to-order approach. The difficulty in scheduling operations in a job shop stems from the wide range of goods, each with their own unique specifications and production routes. When compared to other production settings, such as a flow shop, the required materials, machinery, equipment, and artisanship vary significantly across jobs. Note, however, that the idea of a product route is distinct from that of a task sequence. The sequence of procedures in which a product must visit distinct activities to change inputs into outputs is referred to as routing. The sequence of jobs, on the other hand, refers to the order of jobs entering a work center to be processed. A scheduling activity also involves the sequencing of jobs arriving at a work center. In other words, the scheduler figures out the order of processing for a group of tasks at the corresponding work center. Like forward and backward passes in project scheduling, the scheduler can assign projects to the corresponding work center using either forward or reverse scheduling.

Operations planning is a design or system technique to continuously assess whether supply and demand are in balance. A scheduler should be able to generate reliable schedules by gathering information from both the demand and supply sides of the shop floor. For instance, demand data may be obtained via orders, while supply data can be obtained from work centers on the shop floor. It is important to note that a planner must pay particular attention to bottlenecks since alleviating a bottleneck enhances the efficiency of operations by improving the speed of flow, which is directly in relation to the cash flow from operations.

Planners can use shop floor control technologies to turn data into schedule information. Status and exception reports are often used shop floor management tools that include, but are not limited to, dispatch lists, delay reports, performance summary reports, and shortage lists. Other examples of these control tools include input-output control lists as well as Gantt load charts for work centers. Furthermore, since incoming projects might vary, separate activities can need varied setup or material handling times, which also must be incorporated into an updated schedule. Heizer et al.

(2017) and Jacobs and Chase (2018) supply further information on these technologies. There are diverse types of computer software systems, such as manufacturing execution systems, which take data from shop floors and use it to generate reports that may be used to check and manage activities on shop floors.

It is critical for a planner not just to use computerized methods to help with scheduling, but also to understand the reasoning behind sequencing and scheduling since a company's operational goals might very well change from time to time. For a certain planning horizon, it might be critical to complete a task by its due date but reducing the amount of work-in-progress inventory must be prioritized in the face of systemic risk. Other scheduling aims should include decreasing a task's lead time or finishing a job before its due date. It might be impossible to meet a pre-determined schedule aim, and conflicts can occur if all goals are tried at the same time. A planner, for example, should evaluate how the use of production factors leads to higher levels of work-in-process inventory. Finally, a well-known rule of thumb for a scheduler is to *not* interrupt an existing operation unless it is unforeseen, such as a machine malfunction or an absent worker (Chase et al., 2006).

According to the structure of a manufacturing environment, there are four distinct scheduling scenarios:

- Scheduling n number of jobs on a single work center
- Scheduling n number of jobs on two work centers
- Scheduling n number of jobs on m number of work centers
- Scheduling n number of jobs on n number of work centers

To construct a proper operations' schedule in the last two manufacturing contexts, mathematical models of optimization theory and statistics might be needed. For example, one might choose to use linear programming or simulation approaches to design a schedule for n number of jobs on n number of work centers. Scheduling n number of tasks on m number of work centers, on the other hand, might not be computable for numbers over a certain limit since there will be $n!^m$ different schedules from which to assess the best schedule. Simulation approaches might be explored in such circumstances.

N/1 problem

The N/1 problem describes the simplest form of an operations scheduling challenge, which is to figure out the order in which a collection of tasks must be completed on a single work center. There is a wide variety of algorithms, each of which might offer varying solutions. These kinds of algorithms are referred to as priority rules in operations planning research literature. The following is a list of examples of these priority rules or sequencing heuristics:

- First-come, first-served (FCFS)
- Last come, first served (LCFS)
- Shortest processing time (SPT)
- Longest processing time (LPT)
- Earliest due date
- Random

Priority rules order the tasks to be given to a work center. In that sense, it is like a precedence relationship that defines the order in which works given to a work center are managed.

The titles of priority rules are explanatory in and of themselves. For instance, the priority rule of FCFS ensures that the tasks at a work center are completed in the order in which they were received by the department. A task or a client has the greatest priority and is processed first if they are the first ones to arrive at the location where they are to be managed. Orders in LCFS are arranged in such a manner that the most recently arrived task is overseen first. In SPT, tasks are arranged by processing time, and the ones with the lowest processing times are at the front of the queue. In the LPT rule, the tasks with the longest processing durations are selected to be processed first. It is the exact reverse of how the SPT rule works. EDD prioritizes tasks based on their due dates, thus the job with the earliest due date is run first. Finally, a random order is analogous to a planner acting on a whim, where tasks are organized without regard for rules.

A planner must be able to evaluate the ability of the generated schedule to fulfill a set of criteria to assess the performance of these priority rules on the scheduling activity and to pick a priority rule for schedule preparation. The following are often used criteria for picking a schedule from those generated by the priority rules:

- Meeting due dates
- Minimizing the time that a job spends in the process (the flow time)
- Minimizing work-in-process inventory
- Minimizing worker and machine idle time

After constructing the sequence of tasks that arrive at the related work center, each schedule is compared to the others in terms of their competence in these performance criteria.

Scheduling has numerous and substantial applications across a wide range of industries. Operations must be properly planned, and responsibilities must be delegated in accordance with that planning, to lower supply chain costs overall. In the operations management and research literature, there are many introductory examples of scheduling a work center (Tayalı, 2021, 2012, 2016).

Example 10

Consider a work center with four jobs waiting to be processed. Jobs' processing times (duration) and due dates (days hence) are given in Table 13. Note that the order of arrival of jobs are same as their alphabetical order in this example. The total number of alternative schedules is 4! and to decide heuristically on the processing sequence of these jobs, a planner might choose among various priority rules to evaluate each schedule. The schedules' ability to minimize flow time and lateness are going to be used for evaluating the scheduling performance.

Table 13. Data for work center scheduling (N/1) example.

Job	Duration (days)	Due date
A	4	6
B	5	7
C	3	8
D	7	10

The first alternative schedule is created by the FCFS rule and is shown in Table 14.

Table 14. FCFS schedule for N/1 example.

Pos. No.	Job	Duration (days)	Due date	Flow (days)	Late (days)
1	A	4	6	4	0
2	B	5	7	9	2
3	C	3	8	12	4
4	D	7	10	19	9
			Total	44	15
			Avg.	11	3.8

The position number of jobs, or their order has not changed since the order of arrival is in alphabetical order. The flow time, say, for job C is 12 days (4 + 5 + 3), because it must stay within the system until the work center finishes processing jobs A and B, as well as C itself. On the other hand, the other scheduling performance criterion, the lateness of a job is calculated by subtracting a job's due date from the flow time. For instance, job D is in the system until the end of day 19, but it is due for day 10. Therefore, job D is late by 9 days. In summary, Table 14 shows that the total flow time of the 5 jobs is 44 days, the average flow time for these 5 jobs is 11 days, the total lateness for the 5 jobs is 15 days, and the average lateness of the operation for 5 jobs is 3.8 days. This schedule and its resulting performance metrics are the same for a second alternative schedule that can be created by the EDD rule because the sequence of jobs is the same as that of the FCFS.

The third alternative schedule is created by the SPT and is shown in Table 15. Note the difference in the sequence of jobs which has been adjusted by the SPT priority rule: The first job is the one with the shortest processing time, and the second is the one with the second shortest processing time. This is valid for all remaining jobs.

Table 15. SPT schedule for N/1 example.

No.	Job	Duration	Due	Flow	Late
1.	C	3	8	3	0
2.	A	4	6	7	1
3.	B	5	7	12	5
4.	D	7	10	19	9
			Total	41	15
			Avg.	10.3	3.8

The fourth alternative schedule by the LCFS rule is shown in Table 16 and it sequences the jobs in the opposite direction of the FCFS rule.

Table 16. LCFS schedule for N/1 example.

No.	Job	Duration	Due	Flow	Late
1.	D	7	10	7	0
2.	C	3	8	10	2
3.	B	5	7	15	8
4.	A	4	6	19	13
			Total	51	23
			Avg.	12.8	5.8

The fifth alternative schedule is created by the LPT rule and shown in Table 17.

Table 17. LPT schedule for N/1 example.

No.	Job	Duration	Due	Flow	Late
1.	D	7	10	7	0
2.	B	5	7	12	5
3.	A	4	6	16	10
4.	C	3	8	19	11
			Total	54	26
			Avg.	13.5	6.5

Out of all the rest of the possible different job sequences, Table 18 presents a random alternative schedule. Job C is speeded up in this schedule, in other words, is completed just a day before its due date. Expedition might be a goal for an operations scheduler to achieve.

Table 18. A random schedule for N/1 example.

No.	Job	Duration	Due	Flow	Late
1.	A	4	6	4	0
2.	C	3	8	7	-1
3.	D	7	10	14	4
4.	B	5	7	19	12
			Total	44	15
			Avg.	11.0	3.8

Table 19 presents a comparison of the performance metrics of all schedules that were generated by the 5 distinct sequencing algorithms for the purpose of processing 4 tasks that visited a single work center.

Table 19. Comparison of scheduling performance metrics.

Heuristic	Total flow	Total lateness	Average flow	Average lateness
FCFS/EDD	44	15	11	3.8
SPT	41	15	10.3	3.8
LCFS	51	23	12.8	5.8
LPT	54	26	13.5	6.5
Random	44	15	11	3.8

When faced with these 5 different schedules, an operation planner should choose the SPT option whenever possible since it offers the lowest possible lateness times in addition to the minimal average flow. Although the intuition may be to opt for the LPT and complete the tasks that take the longest amount of time first, this is the poorest schedule among the possibilities in terms of the average flow and the lateness criterion.

Example 11

Consider again an operation with 4 tasks. Table 20 provides information on the order date, the processing times, and the due dates of these projects, with the units of measure being days throughout. The date of the order shows the time that has passed since the order was received. For example, task C was received for processing 4 days ago; its due date is 67 days from the start of the schedule; and it will take the operation a total of 14 days to finish processing work C.

Table 20. N/1 data.

Job	Order date	Due date	Process time
A	14	28	24
B	11	26	15
C	4	67	13
D	9	47	9

Table 21 presents a random schedule based on the alphabetical order of jobs. The flow time is found by adding the end date with the order date, and the lateness is found by subtracting the end date from the due date. For jobs that are not late, late time is 0.

Table 21. Alphabetical schedule for N/1 data.

No.	Job	Order date	Due date	Start date	Process time	End date	Flow time	Late time
1.	A	14	28	0	24	24	38	0
2.	B	11	26	24	15	39	50	13
3.	C	4	67	39	13	52	56	0
4.	D	9	47	52	9	61	70	14
						Total	214	27
						Avg.	53.5	6.8

71

Tables 22 and 23 present the FCFS and SPT schedules.

Table 22. FCFS schedule for N/1 data.

No.	Job	Order date	Due date	Start date	Process time	End date	Flow time	Late time
1.	A	14	28	0	24	24	38	0
2.	B	11	26	24	15	39	50	13
3.	D	9	47	39	9	48	57	1
4.	C	4	67	48	13	61	65	0
						Total	210	14
						Avg.	52.5	3.5

Table 23. SPT schedule for N/1 data.

No.	Job	Order date	Due date	Start date	Process time	End date	Flow time	Late time
1.	D	9	47	0	9	9	18	0
2.	C	4	67	9	13	22	26	0
3.	B	11	26	22	15	37	48	11
4.	A	14	28	37	24	61	75	33
						Total	167	44
						Avg.	41.8	11.0

When compared to the FCFS schedule, the SPT plan offers more favorable results in terms of the average flow time; nevertheless, the FCFS schedule offers superior results for the average lateness measurement. In circumstances like these, a planner could think about gathering more information before deciding. It should be noted that there are other metrics available in the literature for analyzing scheduling performance, such as the utilization measure, which is calculated by dividing total processing time by total flow time. It is possible that a planner might choose to make better use of the facility by going with the schedule that has a higher utilization metric.

A final priority rule for the N/1 problem is the critical ratio, an index number that is calculated for each job waiting to be processed at a work center. To compute the critical ratio, a planner first needs to calculate the remaining time for to meet a job's due date, by subtracting the current date from the due date. The critical ratio of the related job is then found by dividing the remaining time for completing a job by the number of remaining available work time of the work center. Smaller values of the critical ratio show that there is less available work time to meet the related job's due date. Hence, jobs with the smaller ratios are sequenced and run first.

N/2 problem

Scheduling n number of jobs on 2 work centers is called the N/2 problem in the operations scheduling literature. Johnson's rule is a sequencing algorithm used for sequencing and scheduling jobs on two work centers and designed for the minimization of the completion time of all jobs. There is an extension of Johnson's algorithm for a three-work center setting as well.

Johnson's algorithm assumes that both work centers can process each job, but the processing times of each job might vary at each work center. Other assumptions might be developed while running the algorithm. For instance, whether a job can be processed on two work centers at the same time or not depends on the manufacturing setting, and so the decision is up to the planner. The algorithm takes off by selecting the job with the shortest processing time. If the selected processing time is that of the first work center, then the planner schedules the job for the first position of the job processing sequence. If not, the job is sequenced for the last position. In case of a tie in processing times, a planner can decide arbitrarily for whether the related job is going to be processed on the first or the second work center. An assigned job is eliminated from the list of jobs after positioning it on the sequence. The same selection process is repeated for all remaining jobs until they are positioned in the sequence. At the last step, the planner prepares the schedule, after the calculations offer a sequence of jobs.

Example 12

Consider an operation processing 4 jobs at two different work centers. Both work centers have the capability to process all these 4 jobs, but they cannot process the same job concurrently. Table 24 shows the processing times of each job at each work center in minutes.

Table 24. Data for N/2 sequencing example.

Job	Process Time at Work Center 1	Process Time at Work Center 2
A	4	3
B	7	9
C	6	7
D	8	5

Johnson algorithm figures out the sequence by positioning the job with the shortest processing time first. Imagine that there are 4 slots to be filled for creating the sequence. According to the algorithm job A will be positioned at the far right and the last slot of the sequence because it has the shortest operation time at the second machine. If a job's process time were the shortest in work center 1, then according to the algorithm, the planner must have positioned it for the first slot and the far left of the sequence. Image 14 shows the current sequence.

Image 14. Sequencing with Johnson algorithm: Slot 1.

There are 3 jobs left to sequence after placing the A job in its slot. Out of the remaining 3 jobs, job D with 5 minutes of processing time at work center 2, again, gets its position on the far-right side, and to the left of the job A. Image 15 shows the updated sequence with two jobs, and the other two jobs' sequence are unknown yet.

Image 15. Slot 2.

The third job to sequence is, other than A or D, is C. It is sequenced in the far-left slot since it has the shortest of the remaining process times and is at the first work center. Eventually, the final processing sequence of jobs is C-B-D-A, after repeating the steps of the algorithm until all jobs are sequenced. Image 16 shows the final sequence of jobs.

$$\boxed{C}\boxed{B}\boxed{D}\boxed{A}$$

Image 16. Final sequence.

Table 25 presents the final schedule with the C-B-D-A sequence. This sequence supplies the minimum flow time, which is 30 minutes. A planner might also consider using a Gantt chart to visualize this schedule for convenience. However, a schedule like the one provided in Table 25 is satisfactory enough for planning and reporting purposes.

The schedule on Table 25 should be read as follows: Start processing job C at work center 1 beginning at 0th minute, and finishing job C at the 6th minute, because it takes 6 minutes to process job C at work center 1. Then take job C to work center 2 to start processing it at the 6th minute since it was completed

Table 25. Final schedule for N/2 problem.

Job	Center 1		Idle Time	Center 2	
	Start	End		Start	End
C	0	6	0	6	13
B	6	13	0	13	22
D	13	21	1	22	27
A	21	25	2	27	30

at 6th minute at work center 1. Note that there are multiple assumptions in this case. The first one is that the material handling and setup times are negligible. The second assumption is that center 2 cannot start processing job C unless job C finishes at center 1. Work center 2 finishes processing job C at the 13th minute. Work center 1 starts processing job B as job C finishes at the 6th minute. It takes 7 minutes to process job B at work center 1, and just as job C finishes at work center 2 on the 13th minute, job B moves to work center 2. It takes 9 minutes to process job B at work center 2. Work center 1 starts processing job B at the 13th minute, ends it just at the beginning of the 21st minute, but job D waits for a minute since work center 2 processes job B until the end of the 21st minute. Work center 2 starts processing job D at the 22nd minute and finishes at the 27th minute. Meanwhile work center 1 starts processing job A at the 21st minute, after finishing job D. Job A is completed at the 25th minute in work center 1 and it must wait for 2 minutes until work center 2 finishes processing job D. All jobs are finished by 30th minute, and the total idle time of the jobs is 3 minutes. Note that apart from the idle times of the jobs, work centers might be idle as well. For instance, work center 2 is idle at the beginning, waiting for work center 1 to finish job C, due to the assumption of this sequencing problem which is that the work centers are not allowed to process the same jobs concurrently. Next example is free of this assumption.

Example 13

Consider two work centers processing five jobs. Both have the capability to process all these jobs, and they can process the same job concurrently. In other words, a job can be processed on 2 work centers simultaneously. Table 26 presents the processing times of each job at each work center in minutes.

Table 26. Data for N/2 sequencing example.

Job	Process Time at Work Center 1	Process Time at Work Center 2
A	5	11
B	2	6
C	17	8
D	14	13
E	15	7

According to Johnson's algorithm, the resulting sequence for this set of jobs is B-A-D-C-E. Table 27 presents the schedule for this sequence. There is no delay in this schedule due to the relaxed assumption.

Table 27. Final schedule for N/2 problem.

Job	Center 1 Start	Center 1 End	Center 2 Start	Center 2 End
B	0	2	0	6
A	2	7	6	17
D	7	21	17	30
C	21	38	30	38
E	38	53	38	45

77

Next chapter presents a widely used method in operations planning, called linear programming.

References

Chase, R., Robert, F. and Nicholas, J. (2006). *Operations Management for Competitive Advantage* (11th Edn.). McGraw-Hill/Irwin.

Heizer, J., Render, B. and Munson, C. (2017). *Operations Management: Sustainability and Supply Chain Management* (12th Edn.). Pearson Education Limited.

Jacobs, F.R. and Chase, R.B. (2018). *Operations and Supply Chain Management* (15th Edn.). McGraw-Hill/Irwin.

Tayalı. (2021). Manufacturing scheduling strategy for digital enterprise transformation. pp. 104–119. *In*: K. Sandhu (ed.). *Emerging Challenges, Solutions, and Best Practices for Digital Enterprise Transformation.* IGI Global. https://www.igi-global.com/chapter/manufacturing-scheduling-strategy-for-digital-enterprise-transformation/275703.

Tayalı, H.A. (2012). *Time Windows Scheduling Optimization in Low Volume Job Shops* [Istanbul University]. https://acikbilim.yok.gov.tr/handle/20.500.12812/178678.

Tayalı, H.A. (2016). A literature review on production scheduling with the drum-buffer-rope technique. *16th International Symposium for Production Research*, 1085–1090. https://papers.ssrn.com/sol3/papers.cfm?abstract_id=2919846.

CHAPTER 7
Linear Programming in Operations Planning

The field of operations research uses mathematics and statistics to examine any problems that may arise where operations take place, such as in a manufacturing environment. In general, the managers of these environments have two primary fiscal goals: minimizing costs and maximizing profits. To put it differently, a rational entrepreneur should constantly look for ways to deploy finite resources in the most efficient manner while creating products or services.

Optimization theory is a branch of mathematics dealing with the critical values of functions, such as optimum, extremum, or saddle points or values. Cost minimization and profit maximization are common goals of linear programming, which is an optimization technique often employed in the business world for a broad variety of planning challenges. A linear program has solely linear functions and constraints of this model often relate to the available resources of a company. For instance, a linear optimization model may be developed to figure out the ideal mix of production levels for each product to maximize total profit or reduce total cost.

Mathematical models of cost reduction and profit maximization are not just opposites or reciprocals of each other, and a planner must develop distinct models for each sort of optimization issue. Furthermore, optimization problems arise not only in operations planning but also in the field of economics.

This chapter explains how to convert a manufacturing case into the language of mathematics by describing formal connections between decision variables as a linear optimization model, as well as how to code and solve this translated problem using the R programming language.

In economics, one of the most important ideas is the notion of resource allocation. In resource allocation, mathematical techniques and models are used. These methods and models are not reliant on any macroeconomic unit, organization, or ideology. In addition, economists look to approaches for resolving difficulties with resource allocation as a potential rescuer for developing nations struggling with income inequality. In contrast, in businesses, resource allocation choices are seen as intermediate-range and tactical in nature, since they are not as irrevocable as long-term and strategic decisions.

Mathematical programming is an area of mathematics that focuses on the representation of best resource allocation under a set of constraints. A planner must make a variety of judgments about the allocation of resources to conduct their job effectively. Simply said, these resources, whether they come from labor or capital or any other resource, are factors of production. Constraints, on the other hand, relate to the resources that a company has access to, and since these resources are limited, these restrictions form the constraints of the resource allocation issue.

The term 'programming' can refer to multiple things. Computer programming involves creating instructions for a computer to solve a problem. Another use of the term is in generating a schedule for daily life. Mathematical programming is the same thing as mathematical optimization, which involves using mathematical models and algorithms to find the maximum or minimum value of a function. There are various optimization methods used in planning operations, such as linear, nonlinear, dynamic, or integer programming (Chiang and Wainwright, 2005; Taha, 2007; Winston, 2004).

An operations planner might face various difficulties that must be modeled and solved using an optimization technique. The product-mix decision issue is a well-known example of this sort of decision problem. The goal of this problem is to find the number of a certain set of goods that need to be produced to either minimize the costs or maximize the profits. It is important to note that cost reduction and profit maximization are not diametrically opposed concepts, and a planner should develop distinct models for each kind of optimization issue. Furthermore, costs and gains do not have to be monetary metrics in optimization. Time, distance, and mass are all quantitative characteristics that might be taken into consideration. A planner may use relevant optimization models to manage a related issue in a variety of domains, including scheduling, allocating, choosing, and transporting, for example.

A mathematical model must be set up before a planner can begin to tackle an optimization issue. A model is a representation of a system, object, or concept with the intent of explaining a fact. It should build links between the components of the issue by using straightforward language or quantitative formulae. This gives rise to the question of what a problem and a solution are. In the context of a choice, a problem is the difference between an ideal and an actual state, while a solution is the method for dropping this discrepancy. A scientific approach to decision making consists of five steps:

- Define the problem, assumptions, and constraints.
- Set up a model.
- Solve the model.
- Analyze the results and confirm the model.
- Choose the best alternative and apply.

Observing is the first step in the scientific method to problem solving. Not only is observation important in defining the aims and restrictions that are successful in reaching the goal, but it is also important in realizing the simplifying assumptions that make the analysis practical. This first step also includes figuring out prospective solutions and alternatives for choices. In the second phase of setting up a model, one must name the decision variables or criteria that characterize the problem's features. A mathematical model may also need parameter values that affect the issue and reflect the data. A parameter has two features in mathematics: it is both a coefficient and a variable. The last step in this second phase is to define mathematical relations and limitations between decision variables and to translate all of this into mathematics. After constructing the model in the earlier stage, the third step brings in finding a solution to the issue by using optimization methods. This stage often involves sensitivity analysis to look at how multiple factors affect the ideal solution. In the fourth phase, the modeler figures out if the analyzed model reflects reality and behaves as predicted by comparing the model's output results to those seen from the actual system. In the last stage, a planner picks and implements the ideal solution.

Creating a model to solve a problem is useful for analyzing and controlling the parts of the system under different conditions, and even for making predictions about the future. However, statistician Box (1976) is renowned for declaring that all models are flawed, only some are helpful. There are various classifications for optimization problems. For example, if the answer to an optimization problem varies over time, it is dynamic. If, on the other hand, the parameters of the optimization problem do not change over time, the problem is referred to as a static optimization problem. If the model has at least one random parameter, the optimization task is stochastic. A linear programming problem or a linear optimization model to solve a problem has only linear functions, but the model of a nonlinear problem has nonlinear functions.

Linear programming is an optimization method that helps in decision-making for planning operations. It involves maximizing or minimizing a linear objective function to find the optimal solution while providing insights into resource values. The variables in a linear programming model are typically subject to constraints, making it a constrained optimization model. In the absence of constraints, the model is said to be unconstrained. The technique allows one to find the optimal values of a linear function subject to constraints expressed as linear equations or inequalities.

If there exist parametric constants of $c_1,...,c_n \in \mathbb{R}$ such that $z = c_1x_1 + \cdots + c_nx_n$, then the function $z: \mathbb{R}^n \to \mathbb{R}$ is called a linear function. This is the objective function of the linear optimization model aimed at either maximizing a quantitative profit or minimizing a quantitative cost. In operations planning, the objective may be to maximize production quantity or profit, while minimizing production costs or time, and note that cost cutting is crucial for increasing efficiency.

Let z: $D \subseteq \mathbb{R}^n \to \mathbb{R}$; for $i = 1,..,m$, g_i: $D \subseteq \mathbb{R}^n \to \mathbb{R}$; and for $j = 1,..,l$, h_j: $D \subseteq \mathbb{R}^n \to \mathbb{R}$ be functions with $z(x_1,..,x_n)$ as the linear objective function, $g_i(x_1,..,x_n) \leq b_i$ as the linear inequality constraints and $h_j(x_1,..,x_n) = r_j$ as the linear constraints. Then, the general form of a mathematical programming or the linear optimization problem, where x stands for the decision variables, is as follows:

$\max z(x_1,..,x_n)$ subject to, or such that, (s. t.)

$$g_1(x_1,..,x_n) \leq b_1$$
$$\vdots$$
$$g_m(x_1,..,x_n) \leq b_m$$
$$h_1(x_1,..,x_n) = r_1$$
$$\vdots$$
$$h_l(x_1,..,x_n) = r_l$$

Let $c \in \mathbb{R}^n$, $b \in \mathbb{R}^m$, $r \in \mathbb{R}^l$, $A \in \mathbb{R}^{m \times n}$, $H \in \mathbb{R}^{l \times n}$ and $x \in \mathbb{R}^n$; then the above linear program can be written as follows:

$$\max z(x_1,...,x_n) = c^T x = c_1 x_1 + \cdots + c_n x_n$$
$$\text{s.t.} \quad a_{11}x_1 + a_{12}x_2 + \cdots + a_{1n}x_n \leq b_1$$
$$a_{21}x_1 + a_{22}x_2 + \cdots + a_{2n}x_n \leq b_2$$
$$\vdots$$
$$a_{m1}x_1 + a_{m2}x_2 + \cdots + a_{mn}x_n \leq b_m$$
$$h_{11}x_1 + a_{12}x_2 + \cdots + h_{1n}x_n = r_1$$
$$h_{21}x_1 + h_{22}x_2 + \cdots + h_{2n}x_n = r_2$$
$$\vdots$$
$$h_{l1}x_1 + h_{l2}x_2 + \cdots + h_{ln}x_n = r_l$$
$$x_1 \geq 0, \ x_2 \geq 0,...,x_n \geq 0$$

The following mathematical model is the equivalent of the above linear optimization model, where $\vec{x}$ is the vector of the decision variables whose optimal values the problem seeks for the matrices of (A) and (H) which are the technological coefficients matrix of the linear optimization model, and the vectors of $\vec{b}$ and $\vec{r}$ represent the available number of resources for the constraints:

$$\text{Max } Z = \vec{c}^{\,T} \vec{x} \quad \text{s.t.}$$
$$A\vec{x} \leq \vec{b}$$
$$H\vec{x} = \vec{r}$$
$$x \geq 0$$

All functions in this model are linear, which is why these models are referred to as linear optimization models. The matrix of technological coefficients, A, shows the degree of the effect of the decision variables on the resources, defined by the right-hand side coefficients, $\vec{b}$, and that is why A is called the technological coefficients. The solution to this model and problem results in the optimal values for the decision variables as well as the best value of the objective function, and hence the state of the constraints. Readers may turn to the classic authored by Chiang and Wainwright (2005) for more thorough information on themes relating to linear algebra. For the sake of this book's introductory purposes, one should keep in mind that a vector or a matrix may be regarded simply as a table of numbers, while having a deeper geometric basis and a meaning at the same time. Take notice of the difference in notation of the inequality restrictions, which plays a key part in the problem's solution, as well as the direction of the inequality.

The next thing that must be done is to solve this linear model by figuring out the best possible values for the decision variables.

It is common to have multiple decision variables in problems related to operations planning. The simplex method is the most popular algorithm used to solve linear optimization models (Dantzig, 1990). There have been more methods developed. Taha (2007) gives further information on the simplex approach and its modifications for solving linear optimization problems. Various computer programming languages and software incorporate solver algorithms for solving linear optimization models into their software packages. As a result, planners do not need to have knowledge of the engineering of the algorithms to solve linear optimization problems using these languages and software. A production planner must, however, be capable of understanding the manufacturing scenario, translating it into formal mathematics, and then finally into the language used to program computers. The below illustration shows how to first convert a manufacturing scenario into the language of mathematics by constructing a linear optimization model from it.

Example 14

Consider a company that produces two distinct items that flow through four separate work centers. Table 28 shows the expected or predicted process times in minutes for each product. Each work center has 4200, 4500, 4500, and 3600 minutes of available time each week, with unit revenues of $20 and $9 per unit, respectively. Again, the demand for the items is 50 units per week and 110 units per week, respectively.

Table 28. Data for maximization problem.

	Work Center 1	Work Center 2	Work Center 3	Work Center 4
Product 1	35	50	40	25
Product 2	20	20	25	15

A linear programming model can be devised to find the optimal combination of the amount of production of each product that maximizes the total profit. The first step is to assign the decision variables for each product within the vector of decision variables; $x = [x_1 \ x_2]$. The vector of parametric constants of unit profit is written as $c = [20 \ 9]$, or in other words, the coefficients of the decision variables in the objective function are 20 and 9, respectively. Therefore, the objective function is as follows: Max $Z = 20x_1 + 9x_2$. The contribution of producing an x_1 to the total profit is \$20/unit, while producing an x_2 adds 9\$/unit to the objective function. The aim is to maximize this contribution with the set of constraints within the production environment.

The first constraint is the first operation with the following function: $g_1(x) = 35x_1 + 20x_2 \leq 4200$. This equation states that x_1 consumes 35 minutes of operation 1 while x_2 consumes 20 minutes. The total available time for operation 1 is 4200 minutes. In other words, this constraint tells the amount that each product spends in operation 1. The direction of the inequality sign explains that the total amount of operation time available for these two products is only 4200 minutes. The processing time for both products can be less than 4200 minutes but cannot exceed 4200 minutes.

The coefficients of the decision variables in the constraining functions generate the technological coefficients matrix that reflects the quantitative status of the production environment. The rest of the constraint equations are as follows:

$$g_2(x) = 50x_1 + 20x_2 \leq 4500$$

$$g_3(x) = 40x_1 + 25x_2 \leq 4500$$

$$g_4(x) = 25x_1 + 15x_2 \leq 3600$$

The constraints that limit the weekly demand for both products are as follows:

$$g_5(x) = x_1 \leq 50$$

$$g_6(x) = x_2 \leq 110$$

Finally, the amount of production for each product, in other words, the values of the decision variables should be integer and nonnegative:

$$x_1, x_2 \geq 0$$

As previously said, there are alternative methods that one can approach to model and solve the linear optimization problem. The ability to convert a planning scenario into the language of mathematics, which can then be inputted into a computer to calculate a solution, should be the primary skill set of a production planner. The next part presents the R programming language for encoding linear optimization models for operations planning problems and finding optimum points and values.

R programming language

The term programming, in the context of computing and computers, refers to the act of translating a computational issue from its original formulation into a computer language. A programming language may be used to create algorithms. Computer programming languages tell the computer what algorithms to follow.

In principle, all programming languages have the same computing capability, however owing to differences in syntactic structure and other features, various languages may perform differently in different applications. For example, the programming language of PHP (hypertext preprocessor) may be suitable for web site development, while C language may be useful for controlling data networks.

R is an object-oriented, and high-level programming language that is both free and open source. It particularly shines in the realm of statistical analysis. Originating from the S programming language during the 1980s, it became increasingly popular among professionals researching the intersections of mathematics, statistics, operations, and computing in the early 2000s (Jonathan and Kung-Sik, 2008). Researchers around the world have developed software packages that expand upon the fundamental capabilities of R in a variety of subject areas. The programming language can be downloaded onto a computer from the Comprehensive R Archive Network (CRAN) website. R code is executed through the console of the software. An integrated development environment (IDE) can be used with R to facilitate and increase coding productivity and efficiency.

R packages such as 'lpsolve' and 'Rglpk' are designed to solve linear optimization problems and must be installed before they can be used in R. The R command "install.packages" is used to install any package:

install.packages("lpSolveAPI")

After installing the package, its functions need to be loaded and called into the current environment. This is what the library function does:

library(lpSolveAPI)

From this point onward, the package that is going to solve the linear optimization problem is ready to use. Therefore, a planner just needs to code the mathematical inputs of the linear optimization model into R console. The model in Example 13:

$$\text{Max } Z = 20x_1 + 9x_2 \quad \text{subject to}$$

$$g_1(x) = 35x_1 + 20x_2 \leq 4200$$

$$g_2(x) = 50x_1 + 20x_2 \leq 4500$$

$$g_3(x) = 40x_1 + 25x_2 \leq 4500$$

$$g_4(x) = 25x_1 + 15x_2 \leq 3600$$

$$g_5(x) = x_1 \leq 50$$

$$g_6(x) = x_2 \leq 110$$

With the proper matrix notation, which follows the rules of matrix multiplication, the model can be written as follows:

$$Max\ Z = \begin{bmatrix} 20 \\ 9 \end{bmatrix} [x_1\ x_2] \quad \text{such that}$$

$$\begin{bmatrix} 35 & 20 \\ 50 & 20 \\ 40 & 25 \\ 25 & 15 \\ 1 & 0 \\ 0 & 1 \end{bmatrix} [x_1\ x_2] \leq \begin{bmatrix} 4200 \\ 4500 \\ 4500 \\ 3600 \\ 50 \\ 110 \end{bmatrix}$$

The following snippet of code prepares the optimization problem in Example 14 for solving in R, using lpSolveAPI package:

```
model <- make.lp(0, 2)
lp.control(model, sense = "max")
set.objfn(model, c(20, 9))
add.constraint(model, c(35, 20), "<=", 4200)
add.constraint(model, c(50, 20), "<=", 4500)
add.constraint(model, c(40, 25), "<=", 4500)
add.constraint(model, c(25, 15), "<=", 3600)
add.constraint(model, c(1, 0), "<=", 50)
add.constraint(model, c(0, 1), "<=", 110)
set.type(model,1:2,"integer")
RowNames <- c("Operation 1", "Operation 2","Operation 3",
"Operation 4", "Demand 1", "Demand 2")
ColNames <- c("Product 1", "Product 2")
dimnames(model) <- list(RowNames, ColNames)
```

The code begins by creating a model with no constraints and two decision variables. The name of the model is simply 'model' and selected arbitrarily following the rules of R for naming objects without special characters. The "make.lp" function has two attributes, 'nrow' and 'ncol', to define the number of constraints and decision variables, respectively. "lp.control" function adjusts the nature of the optimization model by setting the 'sense' attribute to state whether the model is a minimization or a maximization model. The coefficients of the decision variables in the objective function are input with line 3, using the "set.objfn" function of the lpSolveAPI package. If the problem was to minimize a value, then the 'sense' attribute must have been set to 'min'. 'c' is a function in R that combines objects. The constraints are added to the model object through lines 4 and 9. Note that just like the structure of the constraint function, whether it is an equality or an inequality, the direction of the inequality is also critical in solving an optimization problem. Similarly, the type of decision variables has an impact on the mathematical derivation of the solution. 'dimnames' function of the package assigns the relational names of the constraints to the rows, and the decision variables to the columns of the generic model structure, positioned in an analogous way to the technological coefficients' matrix. The constructed model is printed out in the console after calling the 'model':

model

```
## Model name:
##               Product 1 Product 2
## Maximize       20        9
## Operation 1    35       20 <= 4200
## Operation 2    50       20 <= 4500
## Operation 3    40       25 <= 4500
## Operation 4    25       15 <= 3600
## Demand 1        1        0 <= 50
## Demand 2        0        1 <= 110
## Kind     Std        Std
## Type     Int        Int
## Upper    Inf        Inf
## Lower     0          0
```

To solve the model, one needs to code the following to get a plain 0 as a result showing that there is a solution for the problem:

```
solve(model)
## [1] 0
```

The following lines of code print out the optimal value of the objective function, the values of the decision variables at the optimal solution, and the satisfied value of the constraints with the optimal values of the decision variables.

get.objective(model)

[1] 1900

get.variables(model)

[1] 50 100

get.constraints(model)

[1] 3750 4500 4500 2750 50 100

In conclusion, the company reaches its maximum profit level producing 50 of the first, and 100 of the second product while following the operational constraints.

Example 15

A company produces 3 products with 2 operations. Table 29 shows the required process times for each product in hours. The cost of producing each product is $4, $9, and $3 per unit, respectively and the work centers have a minimum capacity of 10 and 9 hours.

Table 29. Data for minimization problem.

	Work Center 1	Work Center 2
Product 1	2	1
Product 2	3	2
Product 3	1	4

R code below presents the solution for the minimization problem.

```
model2 <- make.lp(0, 3)
lp.control(model2, sense = "min")
set.objfn(model2, c(4, 9, 3))
add.constraint(model2, c(2, 3, 1), ">=", 10)
add.constraint(model2, c(1, 2, 4), ">=", 9)
set.type(model2, 1:3, "integer")
RowNames <- c("Operation 1", "Operation 2")
ColNames <- c("Product 1", "Product 2", "Product 3")
dimnames(model2) <- list(RowNames, ColNames)
solve(model2)
## [1] 0
get.objective(model2)
## [1] 22
get.variables(model2)
## [1] 4 0 2
get.constraints(model2)
## [1] 10 12
```

The cost of production would be if 4 units of product 1, and 2 units of product 3 are produced. Consequently, work centers 1 and 2 would be running for 10 and 12 hours, respectively.

Examples 14 and 15 show the linear programming methodology's primal models. The notion of duality enables the construction of dual models of primal models. The main difference between the primal and dual models is that the primal model can measure the optimum values of the decision variables, but the dual model can measure the value of the resources, which is known as a Lagrange multiplier or, in the context of economics, shadow prices. These values of the resources are referred to as 'shadow' in economics because they are not immediately accessible at the primal model and have been kept in the background, in the shade, so that they can only be accessed by converting the primal model to the dual model using certain mathematical procedures. The next chapter applies linear programming methods to the aggregate production planning problem.

References

Box, G.E.P. (1976). Science and Statistics. *Journal of the American Statistical Association,* 71(356): 791–799.

Chiang, A. and Wainwright, K. (2005). *Fundamental Methods of Mathematical Economics.* McGraw-Hill/Irwin.

Dantzig, G.B. (1990). Origins of the simplex method. *In: A History of Scientific Computing* (pp. 141–151).

Jonathan, D. and Kung-Sik, C. (2008). *Time Series Analysis with Applications in R.* Springer.

Taha, H.A. (2007). *Operations Research: An Introduction.* Pearson Education International.

Winston, W.L. (2004). *Operations Research/Wayne L. Winston, Jeffrey B. Gooldberg.* (4th Edn.). Brooks/Cole.

CHAPTER 8

Aggregate Operations Planning

Planners need to evaluate production factors such as cost, labor, equipment, materials, and processes while setting up a master production plan. The aggregate operations planning optimization model provides a comprehensive view for the manufacturing environment with the aim of reducing costs and enhancing productivity and efficiency. A strategy for making choices, aggregate operations planning helps a company in figuring out the quantities of various production factors, such as workforce and inventory. Literature on the aggregate operations planning problem shows that it is applicable for a 3–18-month planning horizon, although the time window may be altered to suit any time unit.

Multiple researchers have offered a variety of techniques to aggregate production planning problem based on a company's supply capacity and external demand, since the primary assumption of aggregate operations planning is to match supply and demand. There are quantitative techniques, including graphical and spreadsheet models, for resolving the challenges of aggregate operations planning. A linear optimization adaptation of the aggregate production planning model was developed in the 1960s, igniting a renaissance in managing operations.

The family of aggregate operations planning models aims to match supply and demand within a time range specified by the planning horizon to minimize expenses connected with a company's manufacturing or production activities. The linear optimization model of the aggregate production planning matches the supply and demand within a time range given by the production planning horizon and strives to minimize overall manufacturing expenses. Therefore, aggregate operations planning belongs to the strategic aspect of the planning and control framework, which also includes long- and medium-term external capacity and inventory planning activities, demand forecasting and finance planning.

When developing a strong production plan, planners take cost, personnel, machinery, supplies, and procedures into consideration. The aggregate operations planning model combines these characteristics to give a foundation for sound multiperiod operations planning. In other words, it is a technique for making decisions that helps a company decide the amounts of various production factors, such as labor and inventory, while keeping an aggregate perspective of supply and demand in mind. In general, methods of scheduling and sequencing are employed for short-term horizons of days or weeks, while methods of aggregate operations planning are used for longer-term horizons of months or years. The literature on the aggregate operations planning problem shows that it serves for a 3–18-month planning horizon, but the horizon can be changed to any time unit (Tayalı, 2021). Consequently, aggregate operations planning falls within the strategic part of the planning and control framework (Nahmias and Olsen, 2015), which also encompasses long- and medium-term external capacity and inventory planning activities, demand forecasting, and finance planning.

The family of aggregate operations planning models aim to match supply with demand within a time window defined by the planning horizon, with the goal of minimizing costs associated with the company's manufacturing or production activities. As the primary premise of aggregate operations planning is to match supply and demand, multiple scholars have proposed various approaches to aggregate production planning based on a company's supply capacity and external demand. To do this, a company may adjust its inventory levels, staff size via recruitment or layoffs, production rate via overtime or idle time, subcontracting policies, and usage of part-time employees. On the demand side, a company can try to stimulate sales through discounting during periods of poor demand and delay sales through backordering during periods of high demand. An operations planner is usually in favor of following a hybrid strategy where multiple production factors are managed at the same time.

There are many quantitative approaches, such as graphical and spreadsheet models, to solve the aggregate operations planning problem where the objective is usually cost minimization. To emphasize, these quantitative methods rely on trial and error and are thus prone to inaccuracy (Heizer et al., 2017; Jacobs and Chase, 2018).

In the 1960s, a linear programming version of the aggregate production planning model was created, sparking a revival in operations management. There are studies in the academic literature that propose various approaches to aggregate production planning as well as systematic reviews for a thorough classification of these models (Tayalı, 2021). The linear optimization model of aggregate operations planning presents a multidimensional perspective for the manufacturing environment to reduce costs and improve efficiency. Example 16 presents an aggregate production planning problem.

Example 16

This aggregate optimization problem, derived from Chopra and Meindl (2016), aims to minimize the total costs that incur from the production planning horizon of 6 months. Consider a company producing a specific product with the following demands for the next 6 months period:

$$D_t^T = [800 \quad 1500 \quad 1600 \quad 1900 \quad 1100 \quad 1100]$$

Based on these demand forecasts made for each period in the planning period, the company has 8 decision variables to decide on. The decision variables where $t = 1,..,6$, are defined as W_t, H_t, L_t, P_t, I_t, S_t, C_t and O_t. The linear optimization model includes 8 different production planning variables for 6 months; or a total of 48 decision variables whose optimal values are being examined for minimizing the overall cost of operations within the planning horizon. In other words, the aggregate production planning model aims to decide on these decision variables that figure out the levels of production, workforce, inventory, capacity, and stock out for the demand of the planning horizon of 6 months.

W_t denotes the size of the workforce at month t. The cost of employing a worker for the company is $462 per month since a worker earns $3 per hour and works for 7 hours per day and 22 days per month.

H_t stands for the size of the hired workforce at month t. The cost of hiring a worker is $175 per month for the company and this cost includes expenses and related costs such as education and insurance.

L_t is the decision variable for the size of laid off workforce within month t. The cost of laying off a worker includes expenses such as compensation and is $350 per month.

P_t stands for the volume of production within month t. The cost of material needed to produce one unit is $9 per part and the time needed to produce one unit is 4 hours. This means that each worker can output 40 units a month.

I_t is the decision variable for the number of items at inventory at month t. The cost of holding inventory is $2 per unit per month. For instance, cost of holding 2 units in stock for 3 months is $12.

S_t is the number of items that are out of stock at month t, or in other words, items that are demanded by the customer, but unable to satisfied by the company. Stock out refers to a circumstance in which a client's ordered product or service is unavailable in the supplier's inventory. The cost of not satisfying a unit demanded, or the out-of-stock cost is $5 per unit per month.

C_t is the decision variable related to cost of subcontracted items from an external supplier at month t. It might also be called as outsourcing, the process of issuing or contracting a part of the job, product, or service to an external source, often as an alternative to in-house production. The cost of outsourcing, or subcontracting an item is $24 per unit in this case.

Finally, O_t is the number of overtime hours worked at month t. The wage of a worker for a regular hour is 3$ as told previously. Overtime refers to labor performed by an employee outside of their normal working hours, or the cash received for this work. Naturally, the hourly rate for overtime work is higher than the regular hourly rate and is $5 per hour.

The aggregate operations planning problem aims for the minimum total cost, and so the objective function must include a cost vector, corresponding to the decision variables. Therefore, the optimization problem seeks the minimum of the function Z:

$$\text{Min } Z = c^T x = \begin{bmatrix} 462 & 175 & 350 & 9 & 2 & 5 & 24 & 5 \end{bmatrix} \begin{bmatrix} W_t \\ H_t \\ L_t \\ P_t \\ I_t \\ S_t \\ C_t \\ O_t \end{bmatrix}$$

$$Z_{min} = \sum_{t=1}^{6} 462 W_t + 175 H_t + 350 L_t + 9 P_t + 2 I_t + 5 S_t + 24 C_t + 50_t$$

Subject to the linear objective function, 8 linear constraints of the linear aggregate operations planning optimization model are as follows:

$$g_1(W) = W_0 = 35$$

$$g_2(W, H, L) = W_t = W_{t-1} + H_t - L_t$$

$$g_3(P, W, O) = P_t \leq 40 W_t + \frac{O_t}{4}$$

$$g_4(I) = I_0 = 250$$

$$g_5(I, P, C, D, S) = I_{t-1} + P_t + C_t = D_t + S_{t-1} + I_t - S_t$$

$$g_6(I) = I_6 \geq 200$$

$$g_7(S) = S_0 = 0$$

$$g_8(O) = O_t = 10 W_t$$

The first workforce constraint, $g_1(W)$, shows that at the beginning of the planning horizon, where $t = 0$, the initial workforce size is 35. The second constraint is a balancing set of constraints that equalizes the level of workforce at a given period to the sum of the workforce level of the earlier period with the size of the newly hired workforce at the current period, subtracted by the laid off workforce at the current period. Note that there are 6 constraints for this set of constraints.

The next set of constraints of g_3 is related to the production capacity of the company. It shows that each worker can produce up to 40 units per month at regular time, and 1 unit for every 4 hours of overtime. Therefore, the sum of this amount needs to be less than or equal to the number of units produced in a month of the planning horizon. Note that there are 6 periods and thus 6 constraints within this set of constraints.

The inventory related constraint, g_4, sets out the initial inventory level as 250 units. g_5 is a balancing set of constraints, like the second workforce related constraint mentioned above, which equalizes the inventory levels for each period as the levels of inventory, production, outsourcing, demand, and stock out might vary. Therefore, the left-hand side of the equation, the supply, must be equal to the right-hand side of the constraint, which is the demand at the corresponding period of the planning horizon. In other words, this set of inventory related constraints guarantees that the supply matches the demand. Again, note that there are 6 periods and thus 6 constraints within this set of inventory balancing constraints.

g_6 points out the policy of the company at the end of the planning horizon which tells that the inventory must be more than or equal to 200 units. g_7 is for making sure that there are no backlogs at the beginning of the planning horizon. The last constraint, g_8, is related to the legal overtime limit that the company must follow and requires that the workforce works no more than 10 hours of overtime per month. Note that this is a set of 6 constraints as well for it spans the planning horizon like the previous set of constraints of g_1, g_2, g_3, and g_5.

In addition to all the above-mentioned 26 constraints, the decision variables need to be non-negative. Furthermore, for the sake of simplicity, all decision variables are continuous. This model can be solved using a linear programming solver for R programming language, called Rglpk, a free software suite for solving large-scale linear and related programming problems. To install the Rglpk package, one should run the following code within R console first:

install.packages("Rglpk")
library(Rglpk)

The package is ready to use now. To solve the linear problem using this package, one needs to run Rglpk_solve_LP function which requires the following attributes for the solution:

obj: a numeric vector of the objective coefficients
mat: a matrix of technological coefficients
dir: a vector of constraints' directions
rhs: a vector of the resources of constraints
max = a logical parameter to show the optimization direction

Therefore, for this case, one can start with creating the elements of the objective function. Then to create the technological coefficients' matrix, one needs to define the number of rows and columns of this matrix first. The number of rows indicate the number of constraints, and the number of columns of the technological coefficients' matrix are created by combining the column vectors of the decision variables. In short, this matrix has 26 rows and 48 columns. The matrix is sparse since not all decision variables have related coefficients at the constraints. The following snippet of code solves the linear optimization model of aggregate operations planning problem given in Example 16:

```
obj_fun_coef <- c(462, 462, 462, 462, 462, 462, #W
                  175, 175, 175, 175, 175, 175, #H
                  350, 350, 350, 350, 350, 350, #L
                  9, 9, 9, 9, 9, 9, #P
                  2, 2, 2, 2, 2, 2, #I
                  5, 5, 5, 5, 5, 5, #S
                  24, 24, 24, 24, 24, 24, #C
                  5, 5, 5, 5, 5, 5) #O

W1 <- c(1,-1,0,0,0,0,40,0,0,0,0,0,0,0,0,0,0,0,0,0,10,0,0,0,0,0,0)
W2 <- c(0,1,-1,0,0,0,0,40,0,0,0,0,0,0,0,0,0,0,0,0,0,10,0,0,0,0,0)
W3 <- c(0,0,1,-1,0,0,0,0,40,0,0,0,0,0,0,0,0,0,0,0,0,0,10,0,0,0,0)
W4 <- c(0,0,0,1,-1,0,0,0,0,40,0,0,0,0,0,0,0,0,0,0,0,0,0,10,0,0,0)
W5 <- c(0,0,0,0,1,-1,0,0,0,0,40,0,0,0,0,0,0,0,0,0,0,0,0,0,10,0,0)
W6 <- c(0,0,0,0,0,1,0,0,0,0,0,40,0,0,0,0,0,0,0,0,0,0,0,0,0,10,0)

H1 <- c(-1,0,0,0,0,0,0,0,0,0,0,0,0,0,0,0,0,0,0,0,0,0,0,0,0,0,0)
H2 <- c(0,-1,0,0,0,0,0,0,0,0,0,0,0,0,0,0,0,0,0,0,0,0,0,0,0,0,0)
H3 <- c(0,0,-1,0,0,0,0,0,0,0,0,0,0,0,0,0,0,0,0,0,0,0,0,0,0,0,0)
H4 <- c(0,0,0,-1,0,0,0,0,0,0,0,0,0,0,0,0,0,0,0,0,0,0,0,0,0,0,0)
H5 <- c(0,0,0,0,-1,0,0,0,0,0,0,0,0,0,0,0,0,0,0,0,0,0,0,0,0,0,0)
H6 <- c(0,0,0,0,0,-1,0,0,0,0,0,0,0,0,0,0,0,0,0,0,0,0,0,0,0,0,0)

L1 <- c(1,0,0,0,0,0,0,0,0,0,0,0,0,0,0,0,0,0,0,0,0,0,0,0,0,0,0)
L2 <- c(0,1,0,0,0,0,0,0,0,0,0,0,0,0,0,0,0,0,0,0,0,0,0,0,0,0,0)
L3 <- c(0,0,1,0,0,0,0,0,0,0,0,0,0,0,0,0,0,0,0,0,0,0,0,0,0,0,0)
L4 <- c(0,0,0,1,0,0,0,0,0,0,0,0,0,0,0,0,0,0,0,0,0,0,0,0,0,0,0)
L5 <- c(0,0,0,0,1,0,0,0,0,0,0,0,0,0,0,0,0,0,0,0,0,0,0,0,0,0,0)
L6 <- c(0,0,0,0,0,1,0,0,0,0,0,0,0,0,0,0,0,0,0,0,0,0,0,0,0,0,0)
```

P1 <- c(0, 0, 0, 0, 0, 0, -1, 0, 0, 0, 0, 0, 1, 0, 0, 0, 0, 0, 0, 0, 0, 0, 0, 0, 0)
P2 <- c(0, 0, 0, 0, 0, 0, 0, -1, 0, 0, 0, 0, 0, 1, 0, 0, 0, 0, 0, 0, 0, 0, 0, 0, 0, 0)
P3 <- c(0, 0, 0, 0, 0, 0, 0, 0, -1, 0, 0, 0, 0, 0, 1, 0, 0, 0, 0, 0, 0, 0, 0, 0, 0, 0)
P4 <- c(0, 0, 0, 0, 0, 0, 0, 0, 0, -1, 0, 0, 0, 0, 0, 1, 0, 0, 0, 0, 0, 0, 0, 0, 0, 0)
P5 <- c(0, 0, 0, 0, 0, 0, 0, 0, 0, 0, -1, 0, 0, 0, 0, 0, 1, 0, 0, 0, 0, 0, 0, 0, 0, 0)
P6 <-c(0, 0, 0, 0, 0, 0, 0, 0, 0, 0, 0, -1, 0, 0, 0, 0, 0, 1, 0, 0, 0, 0, 0, 0, 0, 0)

I1 <- c(0, 0, 0, 0, 0, 0, 0, 0, 0, 0, 0, 0, 0, -1, 1, 0, 0, 0, 0, 0, 0, 0, 0, 0, 0, 0, 0)
I2 <- c(0, 0, 0, 0, 0, 0, 0, 0, 0, 0, 0, 0, 0, 0, -1, 1, 0, 0, 0, 0, 0, 0, 0, 0, 0, 0, 0)
I3 <- c(0, 0, 0, 0, 0, 0, 0, 0, 0, 0, 0, 0, 0, 0, 0, -1, 1, 0, 0, 0, 0, 0, 0, 0, 0, 0, 0)
I4 <- c(0, 0, 0, 0, 0, 0, 0, 0, 0, 0, 0, 0, 0, 0, 0, 0, -1, 1, 0, 0, 0, 0, 0, 0, 0, 0, 0)
I5 <- c(0, 0, 0, 0, 0, 0, 0, 0, 0, 0, 0, 0, 0, 0, 0, 0, 0, -1, 1, 0, 0, 0, 0, 0, 0, 0, 0)
I6 <- c(0, 0, 0, 0, 0, 0, 0, 0, 0, 0, 0, 0, 0, 0, 0, 0, 0, 0, -1, 1, 0, 0, 0, 0, 0, 0, 0)

S1 <- c(0, 0, 0, 0, 0, 0, 0, 0, 0, 0, 0, 0, 1, -1, 0, 0, 0, 0, 0, 0, 0, 0, 0, 0, 0, 0)
S2 <- c(0, 0, 0, 0, 0, 0, 0, 0, 0, 0, 0, 0, 0, 1, -1, 0, 0, 0, 0, 0, 0, 0, 0, 0, 0, 0)
S3 <- c(0, 0, 0, 0, 0, 0, 0, 0, 0, 0, 0, 0, 0, 0, 1, -1, 0, 0, 0, 0, 0, 0, 0, 0, 0, 0)
S4 <- c(0, 0, 0, 0, 0, 0, 0, 0, 0, 0, 0, 0, 0, 0, 0, 1, -1, 0, 0, 0, 0, 0, 0, 0, 0, 0)
S5 <- c(0, 0, 0, 0, 0, 0, 0, 0, 0, 0, 0, 0, 0, 0, 0, 0, 1, -1, 0, 0, 0, 0, 0, 0, 0, 0)
S6 <- c(0, 0, 0, 0, 0, 0, 0, 0, 0, 0, 0, 0, 0, 0, 0, 0, 0, 1, 0, 0, 0, 0, 0, 0, 0, 1)

C1 <- c(0, 0, 0, 0, 0, 0, 0, 0, 0, 0, 0, 0, 1, 0, 0, 0, 0, 0, 0, 0, 0, 0, 0, 0, 0, 0)
C2 <- c(0, 0, 0, 0, 0, 0, 0, 0, 0, 0, 0, 0, 0, 1, 0, 0, 0, 0, 0, 0, 0, 0, 0, 0, 0, 0)
C3 <- c(0, 0, 0, 0, 0, 0, 0, 0, 0, 0, 0, 0, 0, 0, 1, 0, 0, 0, 0, 0, 0, 0, 0, 0, 0, 0)
C4 <- c(0, 0, 0, 0, 0, 0, 0, 0, 0, 0, 0, 0, 0, 0, 0, 1, 0, 0, 0, 0, 0, 0, 0, 0, 0, 0)
C5 <- c(0, 0, 0, 0, 0, 0, 0, 0, 0, 0, 0, 0, 0, 0, 0, 0, 1, 0, 0, 0, 0, 0, 0, 0, 0, 0)
C6 <- c(0, 0, 0, 0, 0, 0, 0, 0, 0, 0, 0, 0, 0, 0, 0, 0, 0, 1, 0, 0, 0, 0, 0, 0, 0, 0)

O1 <- c(0, 0, 0, 0, 0, 0, 1/4, 0, 0, 0, 0, 0, 0, 0, 0, 0, 0, 0, -1, 0, 0, 0, 0, 0, 0)
O2 <- c(0, 0, 0, 0, 0, 0, 0, 1/4, 0, 0, 0, 0, 0, 0, 0, 0, 0, 0, 0, -1, 0, 0, 0, 0, 0, 0)
O3 <- c(0, 0, 0, 0, 0, 0, 0, 0, 1/4, 0, 0, 0, 0, 0, 0, 0, 0, 0, 0, 0, -1, 0, 0, 0, 0)
O4 <- c(0, 0, 0, 0, 0, 0, 0, 0, 0, 1/4, 0, 0, 0, 0, 0, 0, 0, 0, 0, 0, 0, -1, 0, 0, 0)
O5 <- c(0, 0, 0, 0, 0, 0, 0, 0, 0, 0, 1/4, 0, 0, 0, 0, 0, 0, 0, 0, 0, 0, 0, -1, 0, 0)
O6 <- c(0, 0, 0, 0, 0, 0, 0, 0, 0, 0, 0, 1/4, 0, 0, 0, 0, 0, 0, 0, 0, 0, 0, 0, -1, 0)

```
A <- matrix(c(W1, W2, W3, W4, W5, W6,
        H1, H2, H3, H4, H5, H6,
        L1, L2, L3, L4, L5, L6,
        P1, P2, P3, P4, P5, P6,
        I1, I2, I3, I4, I5, I6,
        S1, S2, S3, S4, S5, S6,
        C1, C2, C3, C4, C5, C6,
        O1, O2, O3, O4, O5, O6),
    nrow = 26)
```

```
colnames(A) <- c("W1", "W2", "W3", "W4", "W5", "W6",
        "H1", "H2", "H3", "H4", "H5", "H6",
        "L1", "L2", "L3", "L4", "L5", "L6",
        "P1", "P2", "P3", "P4", "P5", "P6",
        "I1", "I2", "I3", "I4", "I5", "I6",
        "S1", "S2", "S3", "S4", "S5", "S6",
        "C1", "C2", "C3", "C4", "C5", "C6",
        "O1", "O2", "O3", "O4", "O5", "O6")
```

```
d <- c("==", "==", "==", "==", "==", "==","==", "==", "==",
    "==", "==", "==", "==", "==", "==", "==", "==", "==", "==",
    ">=", ">=", ">=", ">=", ">=", ">=", "==")
```

```
b <- c(35, 0, 0, 0, 0, 0, 0, 0, 0, 0, 0, 0, 550, 1500, 1600, 1900, 1100,
1100, 200, 0, 0, 0, 0, 0, 0, 0)
```

```
Rglpk_solve_LP(obj = obj_fun_coef, mat = A, dir = d, rhs = b,
max = FALSE)
```

Note that in R programming, the code written after the # sign is not read by the computer and used for commenting on the code. As can be followed from the code, the attributes of the solver function have been defined according to the problem's optimization model. The output of the code is as follows:

$optimum
[1] 169197.5
##
$solution
[1] 33.4375 33.4375 33.4375 33.4375 32.5000 32.5000 0.0000
[8] 0.0000 0.0000 0.0000 0.0000 0.0000 1.5625 0.0000
[15] 0.0000 0.0000 0.9375 0.0000 1337.5000 1337.5000 1337.5000
[22] 1337.5000 1300.0000 1300.0000 787.5000 625.0000
362.5000 0.0000
[29] 0.0000 200.0000 0.0000 0.0000 0.0000 200.0000 0.0000
[36] 0.0000 0.0000 0.0000 0.0000 0.0000 0.0000 0.0000
[43] 0.0000 0.0000 0.0000 0.0000 0.0000 0.0000

The total cost of the aggregate operations planning problem is \$169,198 for the following 6 months. Table 30 summarizes the results of the problem according to the above output of the solver by rounding up all values to the nearest integer. The optimal solution of the multiperiod planning problem suggests the optimal values for all the decision variables for all the planning periods.

Table 30. Solution for the aggregate operations planning problem.

Month	W, Workforce size	H, Hired workers	L, Laid-off workers	P, Produced items	I, Inventory size	S, Out of stock	C, Contract items	O, Overtime hours	D, Item demand
0	35	0	0	0	250	0	0	0	0
1	34	0	1	1338	788	0	0	0	800
2	34	0	0	1338	625	0	0	0	1500
3	34	0	0	1338	363	200	0	0	1600
4	34	0	0	1338	0	0	0	0	1900
5	33	0	1	1300	0	0	0	0	1100
6	33	0	0	1300	200	0	0	0	1100

References

Chopra, S. and Meindl, P. (2016). *Supply Chain Management: Global Edition.*

Heizer, J., Render, B. and Munson, C. (2017). *Operations Management: Sustainability and Supply Chain Management* (12th Edn.). Pearson Education Limited.

Jacobs, F.R. and Chase, R.B. (2018). *Operations and Supply Chain Management* (15th Edn.). McGraw-Hill/Irwin.

Nahmias, S. and Olsen, T. (2015). *Production and Operations Analysis* (7th Edn.). Waveland Press.

Tayalı, H.A. (2021). A novel web-based decision support system for aggregate production planning problem. pp. 135–153. *In:* F. Saruchera (ed.). *Advanced Perspectives on Global Industry Transitions and Business Opportunities.* IGI Global. https://www.igi-global.com/chapter/a-novel-web-based-decision-support-system-for-aggregate-production-planning-problem/274913.

Index

Printed in the United States
by Baker & Taylor Publisher Services